BUILDING DEFECT

Mark Whitby

Building Defect
Copyright © 2022 Mark Whitby
ISBN: 979-8-9864918-2-0

All rights reserved. No part of this book may be used or reproduced by any means, graphic, electronic, or mechanical, including photocopying, recording, taping or by any information storage retrieval system without the written permission of the author except in the case of brief quotations embodied in critical articles and reviews.

Because of the dynamic nature of the Internet, any web addresses or links contained in this book may have changed since publication and may no longer be valid. The views expressed in this work are solely those of the author and do not necessarily reflect the views of the publisher, and the publisher hereby disclaims any responsibility for them.

To order additional copy of this book, contact:

2708 Armidale Road Blaxland's Creek
New South Wales 2460 Australia
+1 347 878 1961/+612 8006 0204
Info@shrubspublishing.com

Building Defect

CHAPTER 1 18
    Poor Quality of New House Construction Seems to be the Norm Today 18

CHAPTER 2 29
    Why the Quality of Residential Building Work Has Deteriorated Substantially... and so Fast. 29

CHAPTER 3 45
    Definition of 'Defect' 45

CHAPTER 4 72
    Many Builders and Building Consultants Hide Behind the Word Maintenance. 72

CHAPTER 5 84
    Why Most Building Consultants Find Very Few of The Defects in Your New Homes 84

CHAPTER 6 90
    Before You Use a Building Consultant, Be Sure to Ask What His /Her Disclaimers Are. 90

CHAPTER 7 101
    Time Bomb. Many Buildings Will Fail Prematurely After the Building Warranty Ends. 101

CHAPTER 8 130
    Greed and Privatization of the Building Industry Have Caused Considerable Blight. 130

CHAPTER 9 152
    The Ex-Building Commission Failed to Stop the Declining Standard of Building Work 152

CHAPTER 10 167

BuDomestic Building Warranty Insurance Covers Far Less than You Need It to Cover. 167

CHAPTER 11 174

Builders and Insurers Fight You Because It Pays to. So Many Home Owners Just Give Up 174

CHAPTER 12 183

When in Dispute, Ask the Wrong People and You Will Get the Wrong Answer 183

CHAPTER 13 187

Building Contracts Need Updating Urgently 187

CHAPTER 14 194

Paint Tin. Why Specifications are So Inadequate Today. Does Anyone Care? 194

CHAPTER 15 208

What You Can Do to Give Your New Home a Fighting Chance of Being Built Properly 208

CHAPTER 16 214

What Needs to Be Done RIGHT NOW to Stop Blight in the Domestic Building Industry 214

Forward

I have inspected just over 400 under-construction and recently completed new homes and additions in Victoria, Australia. It is likely that I have seen an atypical set of new homes and additions. But even if the quality of the remainder of the homes built in that period of time... now over two million of them... is just 25% as bad as the poor quality of housing I have seen in Victoria... then we had better do something about it ... and pretty soon.

Perhaps I have only seen the worst side of the total equation. In this sample were my 140 new home owner clients, who already knew that their houses had problems. What they spotted however, was only the tip of the iceberg – there were other hidden problems totalling more than 6 times the cost (on average) to rectify the items that they themselves had discovered.

Anyway, because of my sample being quite small, I have decided not to quantify the problem as I see it to be, concentrating instead on the basic building blocks of quality.

Some recurrent defects are mentioned in the book, because they were spotted 15 or more times in a row. Some defects have appeared in over 120 cases in a row. And these commonplace defects in these new homes had been installed by sub-contractors, supervised by builders or more often by their supervisors, inspected by registered building inspectors (in some cases by the relevant building surveyors themselves); and in many cases also inspected by building consultants (sometimes on multiple occasions by in-house quality control consultants). And yet all of these people missed all of these commonplace defects.

It made me wonder why nobody seemed to find these defects... and why the authorities never mentioned them. It made me think about just how big the overall problem of poor quality building work might be. The problem seems to me to have the potential to be so large, that it may eventually affect the whole economy... and in the not-too-distant future.

So, to open people's eyes, I thought that I would attempt to start a cure to the disease of poor quality from within the industry framework... at its very heart... by defining the word defect: using my areas of expertise relating to new building work, namely: -

- Building Consulting and building inspection reporting... new homes, new additions, homes under construction and established homes (with over 30 years' experience).
- Building contracts, specifications and Domestic Building Warranty Insurance.
- The National Construction Code (formerly the Building Code of Australia) and other related Standards.
- Quality of Building Work related to specification requirements.
- Precedents formed over certain cases where a great deal of fighting occurred. (It is important that

anyone involved in a dispute is aware of pertinent precedents. Learned counsel may be required to uncover this advice).

I also give advice at the end of the book, on what can be done to help the residential building industry cure itself of this now entrenched substantial drop in standards.

(In Victoria, Australia, there have been quite a number of (repressive) changes to the structure of authorities and authority names... plus the addition of a new authority... the DBDRV... which requires every dispute between builders and owners to take place under an appointed Conciliator from this authority before it is permitted to go to VCAT. This authority basically replaces BACV (Building Advice and Conciliation Victoria), but has a role similar to tough Compulsory Conferences in VCAT).

(These changes are as follows: -

- The Building Code of Australia was re-named as The National Construction Code.
- The Victorian Building Authority replaced the Building Commission of Victoria and also partly controls The Plumbing Commission and The Building Practitioners Board.
- Home owners are now required to hire Relevant Building Surveyors.
- Building Warranty Insurance (VMIA) has been made Last-Resort Insurance and only acts in about 3% of cases, after the owners have pursued the builders to bankruptcy.)

Every effort has been made to remove or single out the Australian (State of Victoria) authority names.

However, the failure of these Australian (Victorian) authorities to stop defects arising at an alarming rate becomes apparent as the book progresses. They are inextricably linked to the processes relating to building disputes in Australia.

Acknowledgements

Many thanks to the resolve of the great majority of my clients who fought the good fight, (most of whom came away with substantial amounts of compensation for their poorly built homes), against a system designed to protect enterprise.

This showed me, and may have shown those involved in and presiding over those cases, that some people really wanted to right a wrong; and really cared about their homes.

It also hopefully showed the system that there are houses out there with huge problems deserving of the title LEMONS or SHOCKERS (or DOOZIES as I sometimes call them).

But very little thanks goes to the system; because many of the builders of those shockingly built homes are still building houses today. I see this to be at least partly due to the building dispute forum (VCAT in Victoria, Australia) system of compulsory mediations and the common law practice of gags, which together are very protective of builders. I present this as an acknowledgement because it did give me the will to complete this book, so that home owners will not feel completely alone.

Thanks also to the experts and legal people who helped conquer the negators in the system who were (and still are) prepared to defend the indefensible and fight, simply because, more often than not, and particularly for lesser amounts, it pays to fight ((owing to the fact that (for whatever reason or combination of reasons), many home owners give up and accept paltry amounts)).

Preamble

There seem to be an unbelievable number of homes out there, each with an unbelievable number of costly-to-rectify defects. You see some of these on TV. Well, they are just the tip of the iceberg in my opinion.

Soon after the privatising of building surveyors (and the certifying building inspectors they use) from about 1994, there has been a rapid fall in the quality of work being built: so much so, that basic terms incorporated into building standards are in peril of being completely ignored; even the word *'performance'*... the very basis for the Building Code of Australia (now named the National Construction Code).

It seems to me that recent guides and codes have been written and/or altered, possibly to free up the industry from restrictions: but with the result that enterprise is protected at the expense of fairness; sometimes even at the expense of safety.

Building contracts are deficient in their definitions, with specifications reduced to the extent that they now commonly contain no workmanship clauses at all. New products are being used whilst tradesmen totally ignore specific manufacturer requirements. I mention a few of these to give you the idea that something is basically wrong with the residential building industry. But reporting the woes of the entire residential building industry is not the intent of this book. I think such a role needs to belong to whatever authority replaces the present top building authority (the VBA in Victoria, Australia which in turn replaced the Building Commission) ... both of which needed to change radically in their basic terms of reference.

For starters... attempts to protect negligent businesses at the expense of home owners has to stop being so one-sided. (Precedents already fought out to conclusion in the Supreme Court under the current Building Act and its related Acts may need to be over-ridden by new legislation enacted to safeguard the rights of home owners and define future contracts and fairer insurance).

Insurance should cover far more, and as a result may well cost more. But it can and should be made a whole lot fairer and give home owners of badly built houses the cover they need and expect.

But rather than write an expose on the Residential Building Industry, I wish to start the process of the building industry healing itself... from within.

I wish to help facilitate a fairer playing field from here on; so that current and future home owners can avoid ending up with maintenance-prone homes. Those of you buying or already in possession of a badly built home may also benefit from this book; as long as you get the help of capable hard working caring people plus take heed of existing precedents.

To achieve this fairer playing field, I have concentrated on one vital word in every contract...

DEFECT

Together with its involvement in the world of opinion reports by building consultants.

When experts disagree on the definition of defect there is so much room for argument that unreasonableness becomes rife and legal costs rise dramatically making many give up.

By defining the word defect; (really just an expansion of what is contained in the current legislation... plus a few finer points), we will almost immediately get rid of much of the unfairness in building disputes. Agreement as to what defects actually exist in each home will make disputes so much simpler. The consultants will have to agree. Can you imagine that? (And this is actually one of the aims of VCAT Mediations anyway I believe).

We could then get rid of the very recent Victorian (Australia) DBDRV compulsory mediation forum... supposedly placed there in the way of building consumers advancing to (our Victorian building case court VCAT) where costs can be exorbitant for legal and so-called expert help... not just to the home owners, but also the builders responsible for the defect-riddled houses. This forum was the government's idea of putting an end to the enormous costs being borne by both sides of thousands of disputes (something to the tune of $6 Billion annually just in Victoria, Australia in 2013)

The cheaper outcomes via this head-banging (Victorian, Australia DBDRV) process will be accompanied by a far greater number of defects being band-aided and/or left unrectified in the new homes. The long-term loss is likely to be slightly worse... slightly worse than the recent shocking situation we've been in... probably more than $18 Billion (2013) Australia-wide per year.

What is saved in VCAT fights will be spent in long-term maintenance, courtesy of the conciliators having no definition of defect whatsoever, but really capable of putting off home owners via DBDRV's driving mantra... to expedite as many disputes as possible. (It is not dissimilar to the tactics used in VCAT Compulsory Conferences... the last chance VCAT forum to settle before a full Hearing.

By defining 'defect' officially, all we would then need would be VCAT Mediations and court hearings reduced to sorting out the scope of works necessary to rectify each defect, and the cost to achieve those rectifications.

******

Once DEFECT is thoroughly defined, contracts will become cogent and grey areas will be removed, so that there will be very little room for argument. Builders will quickly realize that the days of hurrying the construction of their houses to completion (whilst ignoring all of the short-changing) are well and truly over... and quality supervision will be of paramount importance.

Naturally, there would initially be many more defects discovered, but most of these will very likely be discovered by the builders themselves before the problems are compounded...

Because supervision would be upgraded and carried out by experienced people with sufficient time to carry out a thorough appraisal at each stage of construction... just as it was in the good old days when pride was an important ingredient of building houses. And this would be of great benefit to all those concerned with building new houses.

Home Owners would once again have well based trust in their builders.

Short-changing would be greatly reduced; and if tradespeople have to do extra work to achieve the desired end result, then perhaps builders should pay tradespeople a little more for that effort. (From here on I will refer to tradespeople as tradesmen if you don't mind).

And there would no longer be the need to bluff and argue with their home-owner-clients.

And from then on, building consultant reports would contain non-compliant with manufacturer specifications defects, unworkmanlike defects and impending defects, (as they should have these last 24 years), rather than just containing minor blemishes and those defective items relating to actual failure.

Then all we would need is random Building Authority spot checks to ensure that adequate supervision is actually happening. But such inspections should only be carried out by very experienced individuals such as capable building consultants (as later defined), registered very experienced (*unlimited class)* builders, and perhaps experienced foremen of apartments... and only those who are prepared to get into and onto the roof and under timber floored houses.

With reports carried out during construction, defects would then not be covered over by subsequent work, making rectification costs far cheaper than they would otherwise have been, had the houses proceeded to completion.

It makes so much sense to me to make adequate supervision compulsory, whether it is by capable building consultants or experienced builders. The number of defects in houses would then drop dramatically.

With adequate supervision there may no longer be a great need for building consultants for completed new houses... or building case lawyers for that matter.

Wouldn't that be something?

Reports carried out by the building authority would of course need to be carried out under a strict definition of the word DEFECT, so that the people will be able to trust those who give final appraisals (in much the same way that people trust their counterparts, the British 'Surveyors').

Otherwise, we will be in a similar predicament to the one we are in right now; and have been for over 24 years... where no-one in any authority or representing any large organization has said a thing about this patently obvious blight.

So, here is my attempt to start a process that should have started well over two decades ago; a process that will

Building Defect

hopefully improve the quality of work being done from now on, one that dramatically reduces rectification costs, and one that hopefully pressures the governments of the day to improve the fairness of contracts and insurance to home owners relating to at least some of the precedents set in higher courts, owing to the steadfast approach of the building authorities not to thoroughly define the word defect.

Chapter 1

Poor Quality of New House Construction Seems to be the Norm Today

I've carried out something like 400 inspections of homes built since 1996 (plus over 50 homes where new additions were carried out in that time period); and I estimate that the average cost to rectify those homes to be about 20% of the original cost to build them. That's a lot of money... but this percentage will (hopefully) be considerably more than the industry average. (And it would want to be)! Over 200 of these homes became the centre of a dispute between the builder (or insurer) and the home owner. Many were lemons.

My sample will be skewed considerably to the high side of what is actually 'average' in the industry, because most of my 120 abovementioned clients already knew that they were the owners of badly built homes. (They had discovered many gross defects themselves).

After contacting the builder, each had experienced such negative responses or such poor-quality rectifications of the defects that they had earlier pointed out to their builders; that they knew they must approach a building consultant or put up with a house of horrors.

And many had to approach more than one consultant to achieve this. Over 10% of my clients who obtained a costed VCAT report had already sacked their first building consultant, discovering extra (missed) defects themselves. But worse still, these sacked consultants had actually failed to discover even a quarter of the defects in those homes.

Many of my clients had great drive because they knew that they had literally 'been shafted' and had nowhere else to turn. At today's prices several of their homes had defects totalling over $200 000 to rectify. Rectification costings to rectify some of the additions jobs were actually greater than was the contract price to build them.

This book aims to reduce the trauma associated with such badly built homes. People should not have to bear such devaluation in their main asset.

These really badly built homes (or '*doozies*' as I usually call them) really astound me... for several reasons.

Firstly it is almost unbelievable to me that the builders (or their supervisors) actually allowed these defects to be incorporated into the houses they were building or adding to, and that the relevant building inspectors, (sometimes even relevant building surveyors themselves), also failed to discover these defects (or maybe worse still, they may have seen them and still allowed the building construction to continue unabated).

Secondly it is also equally unbelievable that in some instances, the building consultants that these owners hired to inspect their recently completed new homes also failed to find MOST, (and in a few cases ALL) of the sizeable defects (that I call blunders) built into these abysmally built homes. Five (5) out of a sample of 21 missed absolutely EVERYTHING.

Thirdly, it astounds me that for many of the home owners, their builders fought tooth and nail to avoid being reasonable in their offers... and used negating consultants who were prepared to distort basic words such as *performance, standard* and *defect* in order to help achieve that end.

And fourthly it is also unbelievable that our Common Law system and Building Authorities have failed to stop the builders of these abysmally built homes from continuing to build.

******

In case you think I'm kidding about abysmal homes, here's a list of some of what I call blunders in just over a 120 new home / new additions sample, where I've been involved to some degree past the initial inspection stage:

SAMPLE LIST OF BLUNDERS IN BADLY BUILT NEW HOMES AND ADDITIONS SINCE 1996

Building Consultants will need to find defects like these or they may well be held to be negligent. And Registered Builders carried out these works and Relevant Building Inspectors approved them. Four of these blunders were actually inspected and approved by Relevant Building Surveyors (Australian classification for those in charge of building permits):

Building Defect

- No footings at all under two 1.5M high brick retaining walls
- No stump pad footings under the stumps of a house (with 60mm fall in the floor)
- No underpinning and no root barrier installed (although drawn on the structural engineer plans and required by the soil report) – several cases
- 3 lower storey posts and a beam missing under upper storey already tiled roof additions by one of the bigger builders
- Garages with inadequate transition zones, where the cars could not enter without severe scraping of exhausts or in one case a garage lintel that the owner's Range Rover could not get under – 3 cases with almost identical situations adjacent
- No party walls in the roof space – fire insurance voided – 3 pairs of units & 9 shops
- Roof tile fixings at below 5% of requirement – 3 cases, one with zero nails / zero clips
- Garage slab (to be by owners) to be laid inside an otherwise completed garage on top of approximately 1.8 metres of un-compacted fill
- A basement workshop with just 1.65M clearance under the beams and also 450mm below ground level left undrained (the photograph of the 1.65M high owner in gumboots was a classic)
- Actual failure of roof trusses (not uncommon and would have worsened) – 2 cases
- Two storey additions inspected by a building surveyor and built on an existing 75mm concrete slab with no edge beam and with under-sized reinforcement ending not even close to the loaded edge
- Very bouncy tiled roof (how do you find this if you don't physically get onto the roof?) due to a severe

lack of bracing and out-of-vertical trusses resulting in double 20mm ocean waves each side of the roof
- Un-drained site with strict soil slip possibilities and embankments double the maximum height permitted by the soil report left un-retained and un-drained

SAMPLE LIST OF BLUNDERS IN NEW HOMES AND ADDITIONS (CONTINUED)

- Grossly under-stated bush-fire risk with resultant inadequate linings and glazing
- Grossly inadequate tie-downs to roof framing of a metal roofed house – 3 cases
- Rotted elevated balustrading and floor joists and bearers at just 4 years old - 2 cases
- 1700mm of soil and a floating garage slab retained by a single brick wall with no agricultural drain (one of 3 major items missed by a previous building consultant working for a larger firm of building consultants)
- Large timber deck over carport with flimsy carport posts on stirrups retaining about 1000mm of soil, not drained and with unsafe carport and deck being used.
- Virtually no sub-floor ventilation (one a 3 year old shop with less than 5% of the total required ventilation with rotted flooring & rotted ends of floor joists – 3 cases, one an HIA runner-up award winner
- Floor tile minor joint cracks and creaking tiles – caused by virtually total foam glue failure
- Powdering paint less than 1 year old due to surfaces being too cold when painted – several cases (the ex-Building Commission very unfairly states that paint

defects are not claimable after just 2 years – no exceptions. Does that seem fair?)
- Leaking aluminium windows - 2 of just 3 weatherboard two storey houses inspected
- Timber deck added to additions – substantial termite attack after just 1 year. And the sub-floor was sprayed to deter termites (a very good case for only physical barriers)
- Timber verandah two storeys high with failed bottom rail fixings to upper balcony balustrade… plus the whole two storey verandah inadequately fixed to the house and posts non-durable oregon facing west and not primed also waiting to rot
- Several windows actually rotted during the warranty period – 5 cases with 3 less than one year old where the windows had been imported and approved for use
- Incorrectly embedded steel beams alongside unseasoned hardwood floor framing – large builder (a repeated blunder I saw 4 times by the same builder) / plus another similar case nearby by a smaller builder
- 60 mm falls in floor but only minor plaster cracks – the fall was mostly there prior to plastering – 2 cases
- Additions with 55mm fall and 16mm cracks – soil report ignored due to computer drawing not being up-graded when the stamped plans showed footing depth of 2.2 metres was required… in red ink – incorrectly diagnosed / monitored by the builder's foundation expert for 2 years with no mention of rectification when this was clearly a serious defect (consultant had not even queried or checked footing depth)
- Grossly Insufficient clearance under the house in a high termite risk area – common problem I believe

- the expense to cure this problem **properly** would be enormous
- A concrete waffle pod slab locally flexing over 20mm in at least 6 locations including where bored piers had been installed – the expense to cure this problem **properly** would also be enormous
- Non-durable weatherboards & external exposed timbers inadequately protected – several weatherboard houses. One 6 years old had rot in over 150 locations
- Steel posts to footings with no allowance for shrinkage of unseasoned hardwood floor framing literally pulling extended house to pieces – 2 cases (1 council-owned)

In three additions jobs, the rectification costs were: for the first 60% of the cost; and for the other two, more than the total cost of the works.

One of these was a classic…

******

The extension was carried out in a richer suburb of Melbourne (Australia), and was a very interesting one because so many costly blunders were associated with just one job, and not a large job at that. In today's terms the contract price would be around the $150 000 or less. However, the following blunders were built into that house:-

- The vertical earth embankment under the house 1650mm high to permit the 2 storey rear additions left just 100mm clear of the lower timber walls, not drained, not retained, without the required

Building Defect

underpinning to the ends to support the overhanging original brick veneer exposed footings, with no protection to the plasterboard lined additions lower storey walls, with inadequate support to original floor bearers and with at least 6 stump pads undermined by the excavation
- The overhanging 2 storey large bay jut-out section (all windows) was built without the required engineer-designed steel corner posts AND the flimsy windows actually extended out past the edge of the lower storey slab by 150mm – so there was nothing for the posts to sit on anyway… and the two storeyed tiled roof additions were being supported by the overhanging aluminium windows
- Under-purlins and rafters undersized / a Barap truss (clever invention using a tightening rod through fulcrums at each end of a clear-span under-purlin forming an upside down truss) missing / 8 collar ties severed to install a new gas ducted heating unit / incorrect propping to an original flimsy hanging beam with a large (unsafe) deflection / inadequate supports (studs & stumps) to the ends of the new propping beams / severe lack of roof batten nails
- A dangerous staircase (one riser less than required and still more problems at the top) resulting in grossly excessive risers and needing complete replacement

BUT there was a final certificate issued for this job… because the (certifying) building inspector hadn't noticed any of these structural defects or safety blunders.

PLUS a building consultant inspecting the works before me found none of these blunders either, but did find over 300 minor paint blemishes (nearly all of them not claimable) that were approved by an insurance loss assessor and previously rectified. That consultant was a painter & decorator I believe, but had been asked to look for all defects in the works.

So I say it's time to do something about these *'doozies'*, the builders who built them, the building inspectors who passed them and the hopeless building consultants who find none or very few of the defects... because 5 out of 22 building consultants found absolutely none of the defects in the homes they inspected; and these defects on average totalled over $80,000 to rectify at todays' prices.

And I wonder why the top building authority... the ex-Building Commission (since replaced by the Victorian Building Authority in Victoria, Australia) has failed to address this growing problem. They seemed not to prosecute builders of 'doozies' very much at all if you read their publications. Why didn't they get the government(s) to audit the industry at least once?

If it wasn't lack of funds, was it because of the privacy afforded to builders in mediations and compulsory conferences? I wonder if the ex-leading building authority in Victoria, Australia was officially told of any of those cases that were actually alleged and often revealed in gory detail in mediations and/or compulsory conferences carried out in VCAT (Victoria's dispute forum in Australia). Was it a requirement that the members keep silent even as to the general malaise so obvious in the industry?

Building Defect

How this enormous problem has remained a virtual secret for over 24 years, without being publicized by the authorities or the larger building consultant firms or their organizations, absolutely flabbergasts me, as it should everyone who reads this book.

So I say there is an urgent need for a book like this in order to start the reversal of what appears to be a very large, costly (and ongoing) blight in the industry and Australia's economy.

And the continuation of this hopeful start to a reform of the Residential Building Industry should be taken up by the VBA... the successor to what was once The Building Commission.

> I'll believe it only when I see it. This message has already been out there for 4 years.

******

Here is a frequently encountered situation at a ceiling hatch where there has been a total lack of consideration for safety of not only the owner but also other tradesmen.

Building Defect

Here the right side support is only supported by the ceiling batten which will not hold the weight of a person attempting to enter the roof cavity… potentially a dangerous situation.

Chapter 2

Why the Quality of Residential Building Work Has Deteriorated Substantially... and so Fast.

Nobody in authority seems to know that there is a growing massive blight in the industry.
The builders, architects, structural engineers, relevant building surveyors, (certifier) building inspectors, even qualified draftspersons and technical specialists, (these people basically comprising most of the pool from which arise building consultants), together with the manufacturers, the authorities and the educators of these people, seem to have had little or no impact on quality control since 1994, when privatised Building Surveyors, the new Building Code of Australia and Private insurance commenced. There were several reasons for this.

The industry seemed to take off and reach boom conditions almost immediately from 1994 onwards. This coincided with easier credit, higher immigration numbers and the rapid rise in mining exports to China in particular. The economy was very buoyant.

The relaxed rules and extra privatisation sped up the permit process dramatically. At first, this seemed to be a great plus for the industry. But unfortunately, there was a down-side.

Gone were the delays caused by letters at fortnightly intervals from councils objecting to a few incorrect items on the drawings. The new breed of private building surveyors realised that speedy service meant repeat builder clients and increased business. Builders could then rush through the documentation stage and even complete some of the documentation such as truss computations after the permit, during the building construction. This seemed good.

But with all the speeding up together with grossly inadequate definitions in the contracts, specifications dwindled to little more than a list of materials to be used in the project. Grab just about any large builder's specification and the chances are that it doesn't contain any workmanship clauses other than a few generalities if that. This means that there has been a complete lack of knowledge as to what is considered workmanlike spreading like a disease throughout the industry for over two decades now. And the fewer people who know what good workmanship entails, the harder it is to keep up standards. This was the beginning of the *dumbing down* process in the residential building industry.

And when the Building Code of Australia (now called The National Construction Code) had useful sections containing basic building methodology removed, standards were again lowered by reducing one of the avenues to adequate knowledge; and so the *dumbing down* process continued. For example when the section on the detailing of floor and wall tiling was removed (as it was in 2004) and replaced in later editions by a lower quality (all in one) sealing process that failed to comply with the requirements of both plasterboard and sealant manufacturers, it's little wonder that more and more showers are leaking today than ever before. And many more will leak in the near future too. This reduction in requirements (with critical dimensions left out) together with an understanding of how showers actually leak, only served to blur what was actually needed by manufacturers, and dumbed down the industry to perhaps its lowest ebb ever.

In the first place, the authorities could, for instance, have insisted on using Villaboard which has the added advantage over plasterboard of not rotting and not collapsing when say cracked tiles are removed. But that would have meant that powerful plasterboard manufacturers would have been disadvantaged. Anyhow plasterboard was permitted in Wet Area Grade form, because it passed a set of stringent tests complying with the code performance rules on the waterproofing of wet areas provided the raw edges of the cut wet area plasterboard sheets were sealed prior to installing with a special paste (seldom carried out it seems). But worse still the new Building Code of Australia details after 2004, showed a greatly reduced gap between the plasterboard and the shower base for instance, with silicone still today shown to be filling a fairly thin gap (not dimensioned as was previously shown as 6mm minimum) between plasterboard sheets and shower base, probably having to pass through some of the glue required to fix the tiles to the plasterboard.

What a joke! But enough said on shower tiling and on to a brief note on roof flashings.

Building Defect

A reasonable attempt was made by the Plumbing Commission when they allowed, for instance, an alternative to the messy time-consuming job of chasing cover flashings into brickwork, but with the provisos that the new system of siliconing the flashing to the brickwork was to have the added requirements of stiffened upper flashing edges, plus the fixings were required at 100mm centres.

Now maybe these requirements were not publicised well enough, but from then on, fixings are now commonly installed about 900mm apart and sometimes considerably more... I guess mainly to save even more time... but also because many tradesmen seem to be deciding what short-changing is acceptable and builders are inadequately supervising that work or simply do not keep up with changes and trust their tradesmen.

This flashing is insufficiently sealed and not fixed to the brickwork, has no stiffened top edge and just render was supposed to seal and cover the gap. The tile also leaves no room for cleaning out the (concealed portion of the flashing called the) under-flashing.

And so the practice of short-changing continues to subsequent jobs until it is basically thought to be acceptable throughout the industry. And then negating consultants call this type of work *standard*, when it follows no standards at all. So why do specifications not stop this in its tracks. I think it is because too few in the industry care anymore.

The roof flashing example was tantamount to *the tail wagging the dog*, because inadequate random drop-in inspections by the authority did not occur to ensure that these changes were being followed.

These are just 2 instances of a multitude of examples I could list, where relaxed rules have created havoc in homes for their owners to sort out at some time into the future. I have yet to see total compliance with either of these examples in any house inspected since 1996. There are many more examples I could have listed with photos to prove them, but the intent of this book is to define defect, so as to make everyone in the industry wake up.

Let's make the residential building industry start afresh as regards what is the required quality in building work and let's make the builders of new homes and additions, the (certifying) building inspectors and building consultants who inspect those homes do a proper job.

Hopefully the definition of defect will also make building warranty insurance fairer... but I fear not until governments first legislate to get rid of what I say are many unfair precedents.

By far the main cause of the drop in standards and quality of work in the Residential Building Industry is the lack of very experienced personnel taking their time supervising the construction of new homes during each stage.

How therefore can new arrivals in the industry in the roles of builder (unlimited), builder (limited), supervisor, architect, structural engineer, building consultant or (certifying) building inspector, fresh out of today's courses, really know what good workmanship is, if there are so many glaring examples of fully approved poor workmanship in the mix that they see. Surely they will need to see a large majority of jobs correctly done to fully grasp what a proper and workmanlike job looks like. Otherwise their experience will count for very little, their opinion as to what is a recognized standard will be warped, confusion will reign, and the industry will become even more blighted than it already is. Thank goodness we hardly ever have tornadoes or earthquakes in Australia.

And so the quality of housing in our largely de-regulated industry dropped to an all-time low quite some time ago in something like just 10 years... and that low standard of workmanship seems to be continuing unabated from there.

Building Defect

Now it is so obvious that the cost to the builder to rectify progressively discovered defects would be very small by comparison to rectifying them at the end of the job.

All the builder would need to do, would be to call the tradesmen back to rectify the short-cuts and other defects they built into the works... for free... before they were compounded by further layers of materials. It could be as simple as that.

Then the tradesmen would think twice before short-changing on their next job. And if the tradesmen were short-cutting because they were being paid too little, then let them fight for more money to achieve what the home owners expect... a workmanlike job.

So what if the builders have to pay a little more in exchange for ensuring that no short-cuts occur. It would be well worth the extra expense in my opinion... for everyone. And the tradesmen would most likely be prepared to do a far better job than they have been doing in commonplace situations where they have virtually been forced to do their work for less and less as a trade-off for a continuity of work. That's how I see the situation.

I also think that the supervisors used by larger builders are over-worked and have to spread their duties so much, that they cannot even get to some sites to assess the workmanship of the tradesmen their builder uses, before the next layer is added to the buildings they are supposed to be looking after. Progress seems to be required at the expense of quality.

I think that there had better be some action soon or we may see a very costly maintenance repair industry blossom... and I wonder how good the quality of that rectification work will be with virtually no rules or standards and the builders in control of that work. It may result in a huge number of short-cuts using materials such as no more gaps, mastic and silicone simply to maintain but not rectify the defects. There will be little thought of the future at all.

So, to all home owners out there... While we have the system that permits builders to employ supervisors, (many of whom are not registered as 'unlimited responsibility builders')... to oversee their house construction; I think there is a great need to obtain your own independent capable building consultant help until the current system is overhauled. But never lose sight of the fact that building consultants vary incredibly in their expertise and level of care. I believe consultants don't even need to be insured for the reports they carry out, and they don't need to be registered building practitioners or architects. Many do not get on or in the roof or under timber floored houses. Some do not inspect metal roofing.

And yet they are prepared to masquerade as building consultants and use the title 'expert'.

Several times I have come across jobs where a succession of supervisors (whilst working for some of the bigger building companies), had to be sacked for ineptitude. That is why I am adamant that there must be a huge increase (at least in the short-term) in the policing of workmanship by the authorities, to ensure that quality supervision does actually occur.

And the companies that fail the test must be made to employ registered 'Builder Unlimited' practitioners to carry out that supervision from then on... otherwise quality (although it has already fallen markedly) may fall even further.

Adequate experienced supervision is just not going to happen until it is required I fear.

Several times in on-site disputes, I have heard supervisors say that the 'builder unlimited' personnel in their companies never even appeared at the sites at all. (Two of my clients have also heard this from the 'builder unlimited' people themselves). So how does the knowledge pass from the top personnel (the *unlimited* responsibility builder - sometimes the sole experienced person) to the supervisors many of whom have insufficient experience and do not know some manufacturer requirements, lack basic building science knowledge, lack knowledge of the Australian Building Code... and are nearly always over-worked.

The answer (I think), is that this experience and knowledge does not get transferred at all. Such knowledge is acquired by seeing what happens by being on site most of the time. It is called supervision because the person doing it closely scrutinizes the work at each stage, before other layers are added. It is not just managing a job and ensuring that the job is completed on time. And it requires a sound knowledge of most aspects of labour.

I think the pre-requisite for becoming a supervisor should be a minimum 3 years as a full time (unlimited) builder in the particular sector of the industry, preferably building a mix of housing types. Then and only then, (as is the case for nearly all professional jobs), can such a person be responsible for the most important task, namely the checking of the quality in the houses they build. Considerable study and/or lengthy on-site experience equips people with sufficient expertise to be an 'authority' on virtually every trade. I say the latter especially is the pre-requisite for supervision.

The builders charge you for this expert supervision, but so often it just doesn't happen. And that is a failure of a duty of care... and that is called negligence I believe.

Experience is just as vital for building consultants; experience obtained by inspecting a large number of a range of house ages and types... especially in inspecting new houses, where I say future life expectancy needs to be considered for every component and every system.

Would it be asking too much of the authorities, to insist that the *Unlimited* Builders who employ supervisors of Builder *Limited* status, actually supervise those supervisors at least half a dozen times during construction or be personally liable? It might cure a lot of problems.

Adequate supervision would virtually ensure that work is carried out in a workmanlike manner as it was about 25 years ago. Poor quality work could be singled out and corrected without too much expense before it was covered up. And disputes would almost vanish.

The law relating to negligence must surely require that supervision (one of the duties of the builder) is carried out adequately... and contracts based on pertinent Act(s) certainly seem to require good workmanship: so by implication supervision must ensure this occurs.

The bigger builders and those who saw that enlarging their enterprise meant bigger profits (via bargaining down the price of materials and labour due to increased volume and virtually continuous work), stretched their experienced expertise very thinly indeed. Most of them basically became far too big to ensure that experienced supervision occurred. Most also chose not to employ builders (of the unlimited type) on a franchise system, with the resultant loss of experience on most projects plus a loss of experience in the industry due to the big builders out-competing the smaller builders on price. It gave bigger builders a huge edge over the now almost extinct new home small builder operators once prolific up until the early 90's.

The smaller builders see out their time doing additions and dual occupancy developments in order to make a reasonable profit. So since 1994, when privatisation increased in the building industry, nearly half a generation of experience has already virtually been lost.

To put it simply there are too many Indians and not enough chiefs.

Unfortunately the authorities have trusted that market forces would force builders to maintain quality control, and they have not carried out a sizeable number of random spot checks to ensure that their trust was founded. They failed to think of the possibility that greed could get the better of pride.

But to do so little about the obvious poor workmanship problem out there... and for so long, those authorities concerned should hang their collective heads in shame for letting this debacle happen. To rely on the fact that only 3.5% or so of owners actually take an action to VCAT is pathetic, and simply shows me that they fail to realize how their own system puts so many home owners off.

Sometimes I wonder... do they care for home owners at all... or just enterprise. Because most home owners SHOULD have taken an action... this is my point.

It is so obvious to me from the ex-Building Commission's regular publication INFORM, that there has been a very sad lack of spot-checking on the quality of builder workmanship by the Building Practitioner's Board for instance, possibly due to the onus being on freeing up the industry and letting market forces take over.

The only check of any significant size that I remember actually hearing about, was in about 1999, where more than 100 new home projects (don't quote me on the actual figure), being built mostly or solely by one large builder, had those projects halted, and those projects made to comply with current (roof) scaffolding regulations. But this was not carried out by the Building Commission via the Building Practitioners Board, but by a Union... for the safety of roof plumbers and roof tilers.

Prosecutions against building practitioners by the Building Practitioner's Board seem to have been oriented towards other types of misdemeanour, where 'complete disregard of the basic rules have occurred. And there ARE quite a lot of these as well.

But there should have been far more prosecutions of the type that exposed the builders of really badly built homes that I call 'doozies'.

Perhaps the Building Practitioner's Board had all their time swallowed up by the blatant types of misdemeanour, rather than the ex-Building Commission expanding and getting people out there to inspect quality of workmanship in houses at various stages of construction. They certainly would have been alarmed, as I have been, at the general malaise so noticeable in (what I can only say appears to be) the majority of houses.

What none of the authorities have ever done though… is to thoroughly define DEFECT.

And I don't really understand why. It would simplify so many things.

Building authorities seem to take the approach that defect cannot be defined except degree by degree in the Supreme Court. That's what I've been told by a barrister and by two ex-Building Commission mediators before mediations actually started.

What they were saying in effect was that the House Contracts Guarantee Act 1995 cannot imply what is a defect, even though it clearly states that breaches of stated warranties are defective work. It doesn't make sense at all. It actually defies logic.

It seems to have also permitted group-cronyism to flourish by permitting great distortions of the word defect and other basic terms used in the building industry codes.

But there are several other reasons why so few home owners have taken a VCAT action.

These include the trust that home owners place in the authorities, their builders, their contracts, their building surveyors and the building inspectors those building surveyors use, plus trust in their building consultant when they think they need one.

There was also off-putting advice from BCAV (Building Advice and Conciliation Victoria) and made even tougher by the new authority compulsory dispute resolution forum... DBDRV, and the ex-Building Commission appointee consultants are instructed to look only at the blemishes home owners allege (all without a definition of defect).

And then there is the general fear of the legal system... not altogether unfounded.

******

Quality has deteriorated at an alarming rate since about 1996... soon after the performance-based Building Code of Australia (since re-named National Construction Code), revised building contracts, revised specifications and the privatization of building surveyors and building inspectors commenced.

Building Defect

By sitting back for just over seventeen years for the new approaches to filter through the Residential Building Industry marketplace, (particularly the freeing up of the regulations and the privatisation of building surveyors & building inspectors from 1994), a problem of *ginormous* proportions has been created...

Contracts, privatization of building surveyors and building inspectors, together with the freeing up of the rules in the industry have generally dumbed-down the Residential Building Industry to the point that the quality of houses, additions and units being produced has never been so poor. There are now too few checks and balances.

Boom times and greed have not only increased the size of the problem but have also accelerated the huge drop in the quality of houses and units and additions being constructed.

If you want to be certain that your house is well supervised, in the short term at least, you will probably have just one recourse left... to obtain the services of a capable and hard-working building consultant; one who actually cares about the quality in your home and its ability to reach a reasonable life-expectancy.

To help achieve this result, your contracts should contain an explanation of what a specification must contain plus contain a definition of the word *defect*.

The next chapter makes a concerted effort to define defect... but there may well be quite a fight before the entrenched system allows this much-distorted word defect to be properly defined.

Chapter 3

Definition of 'Defect'

I think that understanding exactly what constitutes a defect could be the single most important matter in the building industry today. Yet there is so little written on the topic.

If we cannot agree on a definition for *building defect*, then confusing case after confusing case will be heard in the Victorian Civil and Administrative Tribunal (VCAT), where it will be left up to the courts and precedent to decide what is clearly a matter for experts to decide. Unfortunately VCAT members and many building consultants are not building experts.

The Domestic Building Contracts Act 1995 certainly goes a fair way to describe much of what I consider a defect to be, when it describes "defective work shall include at least all implied warranties", some of which are stated in each and every contract. But this seems to leave open the possibility that there may be more reasons for work to be considered defective apart from just '*implied warranties*'. It would certainly include stated warranties.

Later on I will discuss the reticence of the legal fraternity to agree that defective work implies defects.

Some attempt has been made by a few organizations to try to delineate between maintenance items and defect items, but they all fail to adequately define defect.

The Code AS 4349.1 – 1995 – Property Inspections – Residential Buildings… defined minor fault or defect (but stated that it is not realistic to comment on them), rather concentrating on significant defects, which were not defined at all… including in Appendix A, where it again mentioned minor fault or defect as including minor blemishes, corrosion (I assume minor), cracking (I again assume minor), weathering("), general deterioration("), unevenness(") and physical damage(") to materials and finishes. I can only assume that a defect is worse than any of the listed types of minor defect, and that a significant defect is worse again… well 'derr'… and there not a mention of new houses of which there are well over 2 million Australia-wide built since 1996, when short-changing really started in earnest.

There was not one mention of the future, not one mention of compliance, not one mention of workmanship. There was also not a word on the likelihood of failure of short-changed systems or components, reasonable life expectancy or fitness for purpose. So what can building consultants expect to learn from the code supposedly covering their area of work?

Surely people in the industry wrote this standard and its replacement in 2006. And surely new houses (under warranty) existed both times. So as you read on, you'll soon see why these documents were (are) totally useless as codes for the inspection of new homes, new additions or homes still under warranty. I'll reserve my full opinion on these documents and their counterpart code(s) AS 4349.0 and AS 4349.3 for another time.

So is it any wonder that most people don't know what a defect is... including building consultants and building authorities?

Consumer Affairs Victoria: Building & Renovation – Building Definitions, defines 'defects' as being "work that is faulty or contrary to the contents of signed plans". Now the dictionary meaning of *"faulty"* is "defective, imperfect, and blemished". And imperfect, blemished items tend mostly to be of a minor or aesthetic nature, and are often played down by VCAT as being trivialities undeserving of court time... and not even covered by warranty insurance.

The ex-Building Commission's 'Guide to Standards and Tolerances' also supports the inadmissibility of minor imperfections and blemishes. The word *"faulty"* is really just a circuitous way of confusing the issue, (rather than expanding the idea), other than to possibly imply that something is broken or not adequate in some sense.

So why doesn't the big boss Consumer Affairs spell it out? And why is there no mention of time or reasonable life expectancy? (So let's re-examine all of the codes and CSIRO documents superseded or produced since 1994, and build in a time frame for defects).

Works "contrary to the contents of signed plans" is possibly meant (via the word plans) to embrace the contract and the specification for the job... and may therefore imply, (again very circuitously), that work must be carried out in a workmanlike manner and that materials shall be of good quality... but why not spell it out so that everyone knows?

And why not insist that contracts define that specifications actually contain workmanship clauses? Because they don't... and it's pathetic.

Apart from this basic component of building work... LABOUR... that (by the way) forms just under half of the work and total cost of houses: there is not even a mention of life expectancy in the Consumer Affairs definition, and also no mention of the need for strict compliance with manufacturer specifications and expert recommendations of soil reports.

So the Consumer Affairs definition of defect basically states that defect items are defective. But apart from stating the '*bleeding*' obvious, it also fails to delineate between defective work and maintenance items.

Maintenance items to me, means items such as squeaky hinges, lock adjustments, minor gaps opening up due to normal framing shrinkage; and similar items that are no fault of the builder and cannot have been avoided by workmanlike installation. In other words there are no negligence implications. They are basically due to wear and tear. (Equipment wear and tear is also normally included under the definition of maintenance and is included in the warranties of the manufacturers of the equipment).

Building Defect

SO THIS CONSUMER AFFAIRS DEFINITION OF DEFECT IS NOT REALLY A DEFINITION AT ALL.

Because of this total lack in the industry of a proper definition of defect, I feel I needed to go through the motions and reveal that virtually nothing has been done to clear up the meaning of this basic term mentioned in virtually every contract for new homes and additions in the residential building industry. OTHERWISE NOTHING WILL EVER BE DONE.

Obviously any equipment used by the owners after completion and which has its own warranty would normally be exempt from being a defect; but what if its installation left the house unsafe or not fit for purpose. A definition of defect needs to embrace such matters.

The ex-Building Commission document 'Guide to Standards and Tolerances' on non-compliant items to be regarded as defects, is an example of partly defining defect without giving any method of defining what else may also be regarded as a defect. The fact that it was once just 4 pages (admittedly small print) in the days of its founder the Housing Guarantee Fund, but is now something like 64 pages, I think clearly points out that this document is based on shifting goal posts and has always been incomplete and based on a lack of a definition of defect, just as are the pathetic Consumer Affairs and Code for Building Inspections attempts.

Even though time limits are applied to certain components such as paint used in the house, what is again patently missing, (as is the case in the Code in charge of Building

Inspections and in the Consumer Affairs document definitions), is how to assess what is a reasonable life expectancy for all components and systems of a house. And by not including such vital time limits or including assessment techniques to help define what is a reasonable time for a component or system to last, the abovementioned document falls grossly short of a definition for defect and seem to have helped mislead most people in the industry, including the legal fraternity.

And so, the courts have been given a grossly inadequate definition of defect; but still have to decide what alleged defects are actually to be regarded as defects. This situation seems to me ripe for incorrect judgements, with so many grey areas resulting from this shemozzle, that a multitude of barristers and solicitors must be rubbing their hands together in glee, whilst they too have little idea as to what defect really means.

The courts also use one (of possibly several) limiting overlays associated with *warranty*; namely that *rectification costs must be reasonable*. In other words, costs must be in keeping with the value of the loss or likely detriment. This results in consultants attempting to come up with cheaper solutions to rectify the defects. I believe it to be a requirement.

It seems that the reason why the legal fraternity often stated that the Building Code of Australia was not actually law, even though it clearly uses words such as "must be a minimum ...", "shall comply with the standard... or recognised manufacturer fixing requirements" throughout the document. I thought that such words were absolutes.

Building Defect

As required by the Building Act 1993, VCAT looks seriously at alternative methods proposed to rectify agreed defects other than removing / re-building to a recognized standard that the builders were supposed to have done in the first place. Very often, these proposed new methods are really just types of maintenance with a life expectancy often making them not much better than band-aids. That's where the word reasonable really gets going.

BECAUSE THE WORD 'REASONABLE' WORKS BOTH WAYS...

VCAT (as per The Building Act) maintains that the rectification costs must be reasonable: and they try very hard not to award pull-downs. (This is based on a precedent I believe, and used as another overlay). But I say the VCAT 'alternative rectifications must last a reasonable time similar to the expectations of the community (home owners collectively); and that a multiplying factor dependent on a warranty time factor must also apply to any inferior proposed solution that is looked at and/or accepted by VCAT.

For instance when a damp-course is not installed in brickwork where required by the National Construction Code (NCC), and a system is proposed to make the wall damp-proof via liquid impregnation warranted for 10 years, then if the wall is supposed to last at least 50 years then a multiplying factor of 5 must also apply to allow for re-insertion each time it needs replacement. This can often be more expensive than the cost of physically replacing the damp-course.

I say that the word defect should be absolute, and that limits should then be set upon its definition by limiting factors associated with the word warranty, any pertinent legislation and possibly some precedents.

It should NOT be the other way around.

So, because nobody else seems to have had a *fair-dinkum go* at fully defining DEFECT, I am.

******

To start addressing this problem I felt there needed to be a consultation with the Oxford Dictionary. Novel idea, eh?

The dictionary has basically two meanings for '*defect* 'as follows:

1. The absence of something essential to completeness; a lack, a deficiency.
2. A shortcoming, a failing; a fault, an imperfection

The old Housing Guarantee Fund Limited (HGFL) verbal definition was related to me on at least 3 occasions, almost parrot fashion as if it were the law, as *"a defect is that which is not performing or appears likely not to perform as it should."*

It sounds reasonable on first appearances, but I never did agree to this definition fully, and fought on behalf of several of my clients against such thinking. Basically, it seemed too simplistic, particularly when it depended on the words 'perform' and 'likely to perform'.

Building Defect

The HGFL definition word *'appears'* seemed to me to require the failure or commencement of failure of materials or structures by the time of the inspection. There seemed to be no built-in time frame for likely premature rot, likely premature corrosion, movements caused by any of the design forces or damage caused by unusual soil movements unless something had appeared to point to that imminent failure. There also seemed to be a lack of cover for items which did not fully comply with regulations and/or manufacturer requirements, where short-changing had occurred.

There was such a heavy reliance on the word *'perform'*, which has a far wider-reaching definition in the National Construction Code (NCC) than most proponents of this loose HGFL definition were prepared to enforce (based on experience). Perhaps they had not even read the NCC performance conditions or never really tried to understand their full ramifications. Merely passing a comment such as "no failure noted to date" often seemed to condone poor quality alternatives: not *'alternative solutions'* as defined in the National Construction Code the successor to the Building Code of Australia)… but rather a distortion to suit knock-backs by the then sole insurer HGFL.

Later on in the book I look at the misuse of the word *'perform.'*

There seemed to be no chance of the inclusion of items which could quite possibly fail, but which had a likelihood of failing of 49% or less, leaving open the second part of the HGFL definition to individual opinion or even whim.

For instance, a strong wind can hit virtually any house, and this event must be able to be resisted by what has been built. The fact that high winds are largely random in nature is not the point. Wind forces MUST be able to be resisted within stringent limits such as 20 or 50-year return speeds used to avoid this occurring. Risk of winds of this magnitude damaging the structure is totally unacceptable for many reasons including safety, and resultant damage caused by airborne roof debris... sometimes even entire roofs.

So any of you readers or your friends who have not claimed much because your consultant was inept or who had knockbacks from your insurer: I suggest that you call a capable building consultant, (certainly one who does not fail the test I later propose), if you now think there might be a legitimate case for extra items, including some of those items discarded during your dispute... because you may have a very good case indeed. But strict time limits will also apply; denying many home owners of what was once their right. Such is our system.

But get only capable help, including only capable legal advice.

I wonder if many building consultants have thought about this vital word (defect) much: it seems many have never given it a thought. You would expect that organizations would have though. But why hasn't a definition come to light a long time ago?

Building Defect

And this is why so many arguments occur. If they can't agree what a defect is, how can they talk meaningfully about alleged defects? It's such a ridiculous situation.

I've said it before, if the consultants agree, then arguments will then be about how to rectify each agreed defect and how much they will cost to rectify. And legal costs will then be far less. There may then be no need for more than a couple of proper mediations with possibly no hearings at all... wouldn't that be something.

But specification must also be defined and this will be discussed at length later on, so that this most important chapter will be cogent. There is just one thing more.

Before I specifically define defect, I should differentiate between *'incomplete'* and *'defective'* items because of a complication to do with insurance policy niceties involving final completion stage in building contracts.

Contractually, incomplete to me, means only an incompleteness that can still be completed without the complication of involving the removal or re-doing of further works that were subsequently carried out. Complications can turn incomplete items into defects...

******

SO to the definition of defect...

Where I say that...

Building Defect

A DEFECT is one or more of the following, where an item of work:

- **Has failed or is not fit for its purpose**
- **Was not carried out in a proper and workmanlike manner**
- **Was not carried out in accordance with a manufacturer's, structural engineer's soil test requirements (often just recommendations)**
- **Was left incomplete but is now complicated by further works (and more costly to complete)**
- **Will result in an unreasonably reduced life-expectancy of component or system**
- **Is likely to fail and/or perform in accordance with the National Construction Code (successor to the Building Code of Australia) performance provisions and/or deemed to comply provisions - particularly where an alternative solution installer's expertise is (or is shown very likely to be) inexpert**
- **Where under contract, that item is charged for and replaced by an inferior substitute (or not included at all), without a variation.**

And to this definition, restrictions or limiting overlays associated with the word warranty may be superimposed, together with pertinent legislation and pertinent fair precedents.

******

Building Defect

Unfortunately the world is often an unfair place; with consultants such as soil engineers for some unknown reason being permitted to recommend further measures without insisting on them... via recommended works to help avoid rather sizeable movements in the footing systems of a house that could be caused by previously removed trees or existing retained trees, drought conditions and possible present or future drainage problems.

The above definition of defect, or sections of it, may be challenged in VCAT and even higher courts, if defect is to be regarded as secondary to warranty. But I say this is tantamount to the tail wagging the dog and basically idiotic, just as was the old rule of *'give way to the right'* for driving on our roads.

Instead of the necessary money being awarded to rectify a defect in a workmanlike manner, compensation can be awarded in its place, based on the *reasonable cost* limitation(s) in the Building Act.

But on the other side of the equation, the National Construction Code also requires that each component and system in each house performs (including lasting a reasonable amount of time). The warranty period of 6.5 years for contracted building work is insufficient time to fully test many components of a building it should not be forgotten. I will explain.

For instance, a recent (foundation soil) drought lasted from late 1995 until mid-2010; over 14 years in total. So you see... houses built since 1995 were (and are) most of them not FULLY tested (for even 6.5 years) by normal weather with its alternating wet and dry seasons, let alone during their warranty periods.

Withstanding ground movements is usually the most strenuous and far-reaching test most buildings face apart from disasters such as gale force winds; which also have to be designed for throughout Australia for a period of time way past the warranty period. In some areas, cyclones must also be allowed for. Earthquakes fortunately are rare in Australia and do not fall under this design requirement mantle.

So by now you will realise that there will be far more defects awarded under my definition than under the present system that seems to be based on the HGFL 'verbal definition'.

This definition will also do considerable good for the industry. It will make builders increase experienced personnel supervision, which in turn will stop many defects from occurring at all; and more importantly in many instances will stop defects being covered over by subsequent works, with the result that individual defects will be easier to repair and cost far less to rectify. Really badly built homes may become a thing of the past.

AND THAT IS THE AIM OF THIS BOOK… to rid the industry of lemons and drastically cut the number of defects.

Misinterpretation of the words *'perform'*, *'likely'*, and the apparent lack of long-term considerations may well have been responsible for the majority of items being knocked back (as I experienced in all of my cases involving the Housing Guarantee Fund Limited often referred to as the HGFL). And because the current home warranty insurance company VMIA seems to base its approach on that of its long gone predecessor the HGFL, its response also seems harsh, and makes home owners fight to obtain a fair result. Add to this the fact that since the days of the HGFL, insurance cover has been watered down considerably by the then Government following the demise of one of the large insurers FAI (via its HIH arm) in about 2002.

Today domestic building warranty insurance is last resort and owners must first pursue their builder to bankruptcy if need be. This has cut insurer costs by 90% but the premium is a total rip-off and has not reduced. The Auditor General in 2011 and the Ombudsman in 2013 described it as a junk product... But still the government has done nothing to restore fairness.

The swimming pool *'fit-for-purpose'* precedent will add another challenge to each and every alleged defect, and is capable of making it very difficult to win full recompense under what I understand to be the resultant more fine-tuned definition of *'warranty'*. But it may also make it easier to win other alleged defects. At least something has been defined.

BUT, the test of any definition is that it can ride those bumps. That is, IF the authorities actually adopt my definition. They may think that there would be increased risk to the industry as a whole, but they should also realize that claims on new houses being built from that day onwards would be far less, because experienced supervision would have to increase hand-in-hand with the new definition, resulting in a large economic gain in the avoidance of future claims and also dramatically reduced future blight in housing generally.

And IF the definition of defect is NOT adopted by the authorities, we will continue to have a system where barristers strenuously cross-examine experts about something those experts are ill-equipped to talk about... defects. And so dirty tricks, distortions of basic building block concepts, even false statement bluffs and incorrect legal opinion will continue to be used in the majority of cases against the owners' inept experts unable to out-argue the negators.

The result will be a run of owners who end up getting far less than what is necessary to rectify the defects in their homes, sometimes insufficient even to pay their own legal costs. I know this happens, because I have done battle for some of those owners who had previously lost.

These outcomes of this present unfair system are called '*Commercial* Reality' by the negators. But I say it's about time we played ball and moved the goal posts further apart so that the owners can kick goals for the sake of their houses that will otherwise become maintenance nightmares for decades to come. I say it's time to stop the charade of legal precedents ruling the world, and to get down to basics for a fairer outcome not only to the home owners but also the builders and the (hopefully) better-paid tradespeople.

And to the Australians buying houses where the new home owners were forced to give up – get a report done by a capable building consultant (one who is not bound by the Code As 4349.1 or by O.H. & S. rules associated with Workers Compensation Insurance); or you may well have made a very poor decision on your largest asset. And to the building consultants... there are well over 2 million post-1996 houses out there... and inspecting new houses is a completely different ball-game from inspecting older ones.

I should again stress that not all building consultants are capable. In fact, (admittedly just based on a sample of just over 20 reports on the houses I inspected at a later date), many are far from being expert. So, find out the calibre of your consultant before you hire him/her. And I repeat: ASK WHAT DISCLAIMERS THEY HAVE BEFORE YOU HIRE THEM.

My definition is certainly not as simple as that of the HGFL, because it takes into account the complexity of the National Construction Code definition of 'performance' PLUS life expectancy. I also thought it vital to consult the Domestic Building Contracts (DBC) Act 1995 that explains what must (at least) be included in the definition via its own definition of *'defective work'*.

But I see no use whatsoever, despite the intent of the section (b) of that act (which states "a failure to maintain a standard or quality of building work specified in the contract"), when contracts and building surveyors accept specifications that contain no workmanship clauses.

******

The (DBC) Act 1995 defines *defective work* as 'including' that which breaches any of several warranties listed in Section 8 and that which fails to maintain a standard or quality of building work specified in the contract. The word 'including' is very interesting indeed. It says to me, that other reasons for defective work may exist. It is therefore not necessarily a complete definition.

And under Section 8, implied warranties concern 'all domestic building work' and without copying large slabs of the Act, it is of particular interest to note that :-

- 8(a) warrants that the work will be carried out in a 'proper and workmanlike manner and in accordance with the plans and specifications', and
- 8(d) warrants that 'reasonable care and skill' will be used.

Workmanlike ALREADY IMPLIES that reasonable skill and care will be used.

And unfortunately the word specification as it relates to the standard contracts used in the residential building industry is defined and seems not to require any general or workmanship clauses at all. Previously I explained just how the inadequate definition of specification could affect quality of product and your knowledge of what is considered 'standard' (i.e. proper workmanship). It seems to render 8(a) and 8(d) of the contract controlling Act(s) virtually meaningless.

Workmanship and performance also go hand-in-hand: one depends on the other.

The National Construction Code is quite strict in its definitions of performance, stating in no uncertain terms that... (moisture) must be prevented from penetrating / stormwater must avoid causing flooding to... / (structures) must be resistant to (various) forces... and so on.

In the HGFL definition of defect, there seemed to be no room for consideration of unlikely events at all. Because it hasn't fallen down, hasn't rotted, or hasn't cracked by the end of the warranty period, then the words unlikely to perform could then take over. I even think that because of their success rate in knock-backs that they would argue strenuously that virtually the whole building had stood the test of time by the end of the warranty.

Well it hasn't... and it ain't good enough!!!... Because...

The definition of 'performance' has a reasonable life-expectancy time frame built into it.

And reasonable life expectancies vary up to 60 years I believe. So the relevance of *old man time* must be considered for each defect alongside the precedent of *fit for purpose*, fought in court over a swimming pool and a diving pool I believe.

This HGFL definition, or at least its application, seemed so insidious to me, that it played no small part in my determination to write this book. I hope that people will see this book as an attempt to start the cure of a decrepit cronyistic industry with a rapidly spreading blight, rather than simply mouthing off at all and sundry. (Yet I think you'll agree that there is good reason to mouth off at many component parts of the industry).

And by the way, you will have already read that in Victoria at least, we have recently just endured 13.5 years of continuous drought? – and by 'building drought' I mean less than normal soil moisture content in the soil deep down... under the footings... in the 'foundation soil'.

So the idea that the foundations have stood the test of time, even now, for houses built since 1996 is to me still generally quite ludicrous.

Differential moisture in foundation soil (below the bottom of footings) is the major cause of cracking in brickwork: and until foundation soils become fully moistened again, no building built in this drought period can possibly be regarded as *having stood the test of time*. And that is virtually every house built since 1997. And because of this, heave must be considered, but rarely is. Just look at the articulation of the brick veneer houses built in this latest period of time... incorrectly installed in over 99% of those houses... more later on.

So what about the future of all of the houses built during the prolonged drought then?

If ground movements cause cracking; and drought conditions keeps foundation soil fairly constant; then at the end of the Building Warranty period, it cannot reasonably be said that brickwork and other footing-related cracking is greater or less than 50% likely. The word 'likely' was

Building Defect

therefore an inadequacy in the HGFL definition, and any similar word should be very carefully considered before being incorporated into any definition of defect.

It quickly becomes clear, (as I have experienced so many times), that by confining discussions to a very narrowed-down inadequate definition of defect, many claimed defects will be (or have already been) wrongly cast aside. And that is precisely what has happened for quite a few years now. The Housing Guarantee Fund goes back to well before 1996 I believe... SO...

ANY ABBREVIATED DEFINITION OF DEFECT BY ANY AUTHORITY HAS A LOT TO ANSWER FOR.

It must surely be a requirement that the word *'perform'* (and *'performance'*), when used in the residential building industry, must be used strictly in accordance with the book that defines them... the National Construction Code, the 'bible', so to speak, of the building industry.

The Housing Guarantee Fund Limited definition to me clearly ignored the BCA (and today ignores the NCC), based on just one of many possible examples; that of a house roof lifting off one day after the warranty period has ended. When you have wind code rules in place that mention 20, 50, 100-year-return storms, then the building has certainly not stood the test of time at the end of its 6.5 year warranty period. Until any lift-off actually happens, (fortunately not frequently), it is definitely not *likely* that a roof will lift off, but it is possible... if the tie-downs are not adequate. These possible events must be considered, (it states in the performance provisions of the National Construction Code). Otherwise the safety of the public is at risk, where it could (and should) have been avoided by installing sufficient tie-downs.

Lack of tie-downs is not just a structural defect, but a safety issue as well. And it seems likely that quite a considerable number of metal roofed houses are inadequately held down. I know of 2 cases in Melbourne where roofs came off in high winds since 1996... because winds are getting stronger... because temperatures are rising, the reason for this being... call it what you like. And roof tie-downs are more stringent than in the past I believe. And timbers being used are softer so that fixings need to be the correct type and need to penetrate further... more on this later too.

So too are most roof ridge tiles inadequately held down by a recently approved product; one used almost exclusively throughout the industry now, because that product is not installed as required by the manufacturers.

The inclusion, by opposition experts, of many defects into the category 'maintenance items' is another misconception (or dirty trick) that denies (or is aimed solely at denying) your rights to have building systems correctly installed in your house construction, so that they last a reasonable time. You can see this mentality in the proposed band-aid solutions proposed by many building consultants at VCAT, instead of adhering to standard (tried and tested by time) workmanlike solutions or those actually intended by the manufacturers of the products abused.

Reasonable life expectancy of the building fabric and systems must be considered.

In fact, a whole host of life expectancies and forces have to be considered under Performance Provisions of the National Construction Code, when appraising the defect status of a new building. I wonder if most experts or authorities have given this matter much if any thought at all.

Most of the reports I have seen are quite deficient in considering reasonable life expectancy. If it's not badly cracked, hasn't fallen off or isn't rotted or corroded, then ignore it, seems to be the general thought process. Building consultants wouldn't be needed at all if that was all there was to it. Most home owners are quite capable of identifying this type of obvious defect for themselves.

Many of the reports I have come across when asked to provide a second opinion, failed to find all of these more obvious defects. Many listed no defects where more than 30 existed. One report failed to list all 55 defects and 22 of the 44 incomplete items. Another missed all 40 items in a 137B report for an owner builder selling a house before it was 4 years old.

Here are some examples of blatantly obvious incomplete works, obvious structural inadequacy or obvious impending failure missed by some of the aforementioned building consultants that I was asked to check up on:-

- Just 2 insulation batts in the roof space
- Over 600 missing roofing screws

- All roof tile fixings missing
- 12 collar ties missing in a conventional tiled roof with props onto hanging beams
- An embankment 2 metres high left un-retained and undrained in a soil slip area
- Wall frames (no kidding) hanging from the roof trusses with 20mm clear gaps under the wall plates/skirtings to the floor that you could roll a marble under
- Over 50 popped nails in a garage plasterboard ceiling 4 years old… after being re-fixed to cure the problem 2 years earlier (but the problem was still there is why it recurred)
- Re-blocking of timber stumps not done to a fully renovated house (2 building consultants one a structural engineer missed this one, and I have heard that firm of structural engineers being recommended by authorities

In the residential building industry there are many authorities such as Governments, Consumer Affairs, the Victorian Building Authority (successor to the Building Commission), the Domestic Building Dispute Resolution Victoria (more repressive successor to Building Advice and Conciliation Victoria), The Housing Industry Association, the Master Builders Association, the Australian Building Codes Board, the Building Regulations Committee, Standards Association of Australia, Plumbing Commission, Building Practitioners Board, Institutes including the Institute of Building Consultants, educators and more, who have or could have had an input into the quality of building work carried out in Victoria, (probably similar to other states).

But with all of these authorities involved in the residential building industry, not one has clearly and thoroughly stated (for new buildings in particular) just what a defect is to date it seems. It's been left to the courts.

If you've made it to the end of this chapter you should see why it is vital to use a building consultant who inspects without blinkers on, and who argues on a full and correct basis.

You should, in my opinion, want to question anyone who has been acting from a blinkered mind-set for some time, to ascertain whether or not they are actually capable of (changing and) adequately representing you... because some building consultants will have a great deal of researching and homework to catch up on if they are going to be able to find things that they have ignored since 1996 (for the past 23 years).

If you have read everything to this point, you may be ahead of many of the so-called building consultant experts out there... and you may need to be for the sake of your homes.

What every one of the building owners I've met want, is for their homes to be well built... built to current standards in a workmanlike manner. They do not want their homes to be built in a manner tradespeople come up with to save time and money... and they want their jobs to be adequately supervised by experienced people who are not over-worked.

That is why building standards exist... to safeguard the quality of work in the Building Industry. This is so that

building will last for a reasonable length of time… or longer, as were the homes built back in the 1960's & 1970's for instance.

No home owner I know wants a house that needs copious quantities of maintenance carried out regularly when those items could have been built to last considerably longer than just a few years.

In other words home owners want their houses and all of their systems and component parts to last what would be considered to be a reasonable length of time; to reach a reasonable life expectancy. So if there is any code for life expectancy, let it be closely examined in the light of my definition and what the public want. (You may well find obvious ignoring of what is actually reasonable).

******

If (certifying) building inspectors and building consultants are repeatedly permitted to fail to uncover all of the reasonably accessible defects, is it little wonder that the quality of work has declined so much?

It is offensive to me that certifying building inspectors either do not bother to find many of the defects in the new houses being built (because they are not required to), or don't even notice them, or worse still, they see them but don't report them.

They are certainly not reporting the large majority of commonplace defects.

So, whichever is the case, they seem to have failed in their duty of care, are therefore likely to be negligent... and should be audited and re-educated as soon as possible.

The same goes for whoever the building consultants are who inspect those houses later on or during construction, because they also fail to find the vast majority of the defects left by the builders of the houses they inspect... and that is all that they are employed to do.

Chapter 4

Many Builders and Building Consultants Hide Behind the Word Maintenance.

Maintenance (discussed earlier) basically means attending to the wear and tear in items that were installed in a workmanlike manner and have lasted as long as they could. But if the frequency of maintenance of certain items that have started to fail is less than reasonable for that system or component part, then there is a negligence factor involved and the maintenance is merely a band-aid and the item is actually a defect.

Short-cuts very often result in unreasonable frequency of maintenance. And many builders take the approach that the maintenance they carry out on what is actually a defect is appropriate, so that one, two, even three maintenance sessions will eventually get them to the end of the warranty period at a far reduced cost to rectifying those defects.

Home owners are then left with systems and component parts that are going to need constant maintenance and could often lead to more costly problems such as rotted floors.

SOME EXAMPLES OF MAINTENANCE TYPE BAND-AID RECTIFICATIONS I'VE COME ACROSS.

Building Defect

1. Silicone re-done across the zero gaps between tiles-to-shower base and corner wall tiled joints.
2. More silicone over failed silicone at the joint formed at flashings-to-brickwork junctions where no cover flashings were installed.
3. Rot in balustrading (only 4 years old mind you) merely silicone filled & acrylic painted to match.
4. Painted fill over popped nails - very noticeable where touched up, and many nails re-popping.
5. Metal corner protectors for timber fascias (likely to become an acceptable band-aid until everyone realises that the rot progresses... eventually past the cover flashing).
6. Roof valley flashings gouged during the valley tile sawing process touched up with pointing final coat.
7. Severed trusses with very short pieces of pine to one side of the resultant gaps held together with only a few nails.
8. No-more-gaps (for the third time) to fill increasing gaps caused by the builder's lack of catering for differential shrinkage of different heights of unseasoned timber floor framing. In two cases, the new gaps I saw were 6mm.
9. The tilting shower base that had a large wedge driven under one side to level it up.
10. The floor tiles replaced in an En Suite that removed the 8mm fall towards the shower base but still left the 14mm fall towards the carpeted bedroom door.
11. The coat of paint over the damaged plasterboard with the window head still leaking.
12. The flashed and silicone-sealed tops of floor joists protruding past the external wall to form the framing for the balcony, turned up the wall behind most of the wall sheeting without removing that wall sheeting and not integral with the window sill flashings... again leaking (as the builder was advised it would).

13. The pushing up of the roof tile battens more than 20mm to keep the tiles planar to compensate for the excessive sag in the roof trusses, thus reducing much of the required inherent bracing of tight-fitting battens.
14. An incorrectly coloured Colorbond roof poorly re-painted in acrylic – 2 coats. The application would be lucky to last 6 years and would then peel, whereas Colorbond coatings last 40 years or so.
15. Flimsy plastic coved skirtings installed around two showers that had leaked because there were no available matching tiles… and a few tiles had to be removed in order to reinstate the showers. (By the way the new showers were again leaking before the warranty period had ended).
16. Silicone over roofing screw grommets that had been forced out by too much tightening.

The balcony tiles after rain squelched water up from between the tiles so the builder came back with a wonder chemical to seal the fine grout cracks in the tiling and this was the result.

Building Defect

This floor had a 15mm hump in it so the builder returned to wedge the remainder of the flooring up to the same level destroyed.

This vanity unit was installed without any tiles and so the builder returned and simply glued the tiles to the wall... to the ordinary plasterboard and with no joint at the bottom of the tiles.

Building Defect

This part-tile used to slip out so the builder got his roof tiler to return to fix it without his supervision and this was the result and the holding dollop of flexible pointing cracked in less than 18 months.

The builder left the brickwork with holes in it and when the owners complained he returned and filled the joints himself with this result.

Building Defect

The builder was asked to return to fix the lack of tape to this and several other black underlay joints **including under the polystyrene foam** at 4.30 PM. He said no worries but the slab was poured at 7.30 AM the following morning.

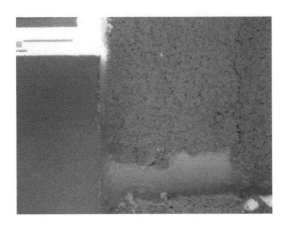

This gap in the render was the result of flaking off....But rather than getting the renderer back the builder simply siliconed under the far-too-thin loose render.

Building Defect

This tile is clearly partly separated from the glue but I was told that it was not going to be replaced as part of the re-tiling maintenance man's repair of the lifted balcony tiles by order of the builder.

Shove the tiles together and silicone the still out-of position under-check chip which will still leak and worse than previously too.

Building Defect

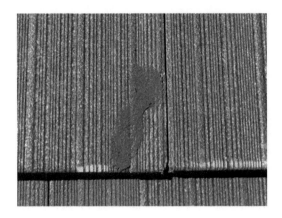

Final coat pointing material smear over quite a long chip will last about 2 years if lucky and then leak.

The builder's maintenance man went FROM THIS…

Building Defect

TO THIS. Dollops of silicone to hold the tiles – about 3 years until the tiles slip again, if lucky… and I bet the dirt wasn't cleaned off first… A mechanical fixing to a batten is required.

After 2 attempts to re-glue the feature mouldings. The manufacturer requires fixings.

Building Defect

There are many more band-aids that I could have listed or pictured. (And by the way, all of these examples had been built and then band-aided at least once by registered builders.

The owners all allowed this to happen. Most of these items were the result of builders saying they would rectify certain defects because the builders could do those particular rectifications cheaply (because this time they had to do it for free). And there was no agreed scope of works for any of these works. I suggest you do not let this happen to your house.

It is because the home owners trusted their builders to do a good job the second time around, when they couldn't do it well the first time, that each owner effectively said to their builder... please band-aid my house. This happened because the owners took the easy path.

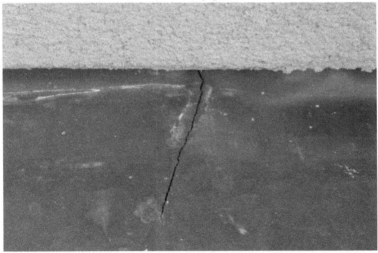

This thin lead flashing (too thin and incorrect type of lead) needed more support... it got siliconed as you would expect... and continues to sag revealing the top of the tear which now leaks. Other tears have also commenced because the inadequate support and fixings.

Many owners choose to take the easier path when they agree to the appointment of a building consultant by a building authority to (virtually) adjudicate on (just) their list of alleged defects... without that consultant having a definition of defect. That is our system in a nutshell. And it's pathetic.

A piece of advice for all home owners in dispute:

Always insist; that if a builder wants to return to the site to rectify the agreed defects; that this is to be conditional upon the builder stating specifically how he proposes to rectify the defects. In other words he must scope the works. Otherwise you will simply be inviting him to band-aid your house. Remember he has already failed to build the works properly whilst being paid, but is now proposing to return to rectify the (easy-to-band-aid) defects... **but this time for nothing (that is the point)**. This is a very reasonable demand considering that you don't have to allow him to return at all. Don't be talked out of this request by anyone. It is one of your basic rights.

******

One novel cure of a hyperbole arch where the base on one side was set 90mm over by mistake was... the sledge hammering back of the wrongly positioned bottom plate... dada! It then just needed a bit of 'No-More-Gaps' and a paint touch-up. When he thought he was not seen, the peg-

Building Defect

leg walk to the car by the builder (seen by both the owner and me) was a *'classic'*.

Chapter 5

Why Most Building Consultants Find Very Few of The Defects in Your New Homes

In order to carry out inspections of new work, be it new homes, 137B reports for owner-builders selling prior to 6 years old, new additions or even new apartments, Building Consultants need the following qualifications and expertise to be called capable:

- They should be registered as a registered building practitioner or registered architect or registered structural engineer (but there is no requirement I believe)
- They should have professional indemnity insurance specifically for doing pre-purchase inspections (many do not and it's not required... again – Why not?)
- They should have carried out reports on at least 500 older homes (many have not)
- They should have carried out reports on at least 25 new homes or additions
- They should be conversant with the Light Timber Framing Code in Australia
- They must be prepared to represent you in court (VCAT in Victoria, Australia) should the need arise
- They should have dealt with experienced VCAT legal people several times

- They should get under timber floors and in roof spaces with adequate clearance
- They must be fit and preferably not too fat so that they can get into the roof space
- They must get on the roof to inspect it adequately (Note: employees do not in Australia for Workers Compensation Insurance reasons)
- They must care about the homes they inspect
- They must not mind getting dirty or climbing into confined spaces
- They must be good researchers
- They must be familiar with a range of manufacturer specifications
- They must have a thorough definition of defect (the vast majority do not)
- They must be prepared to take the time necessary to find virtually all accessible defects (again most do not)

That is a lot of requirements, but I say that a building consultant needs every one of these qualities to be able to call himself (or herself) capable. And it would be good if every building consultant informed owners of their rights should they discover a lengthy list of defects.

Building Consultants fail to find defects particularly in new homes and new additions for a number of reasons, usually including more than one of the following:

- Their lack of effort to unravel what constitutes a defect
- Their lack of building science knowledge
- Their lack of manufacturer specification knowledge
- Their lack of consideration of the future

- Their lack of experience in long-term failure of materials and construction systems
- Their lack of thoroughness (often due to being an employee governed by Workers Compensation Insurance requirements)
- Their lack of inquisitiveness

This has been so clearly displayed in the 22 reports of other building consultants (on the side of home owners) that I have followed so far since the start of 1997.

Apart from most of them appearing to be downright lazy, it was quite obvious that all 22 abovementioned building consultants failed the test for an adequate definition of defect: strange when these so-called experts were looking for nothing but defects.

It was indeed fortunate that I saw these reports, or I may not have kept up the momentum to complete this book.

You see, I assumed that building consultants are mostly knowledgeable people: just as most home buyers and home owners probably do. I now know that this is certainly not the case, and quite possibly the exact opposite is. What a terrible thing to have to write.

Four building consultants missed absolutely every defect in the homes they inspected (31, 34, 42 & 55 defects respectively). Another missed almost 80% of the major defects (9 out of 11 and 3 of these were serious structural defects). All but one (19 of the 20) failed to find even half of the defects, and missed nearly all of the costly-to-rectify ones. Two other reports where (the solicitors say) the building consultants missed all 37 & 41 defects respectively have not yet been sighted.

When you read such reports where none of the defective items (30+ / 40+ / 50+) that you found, including a serious fire risk which would cost $6000 to rectify, rotten balustrading (only 4 years old note and with a permit and occupancy certificate), ceiling and wall insulation totally missing, none of the compulsory tie-downs installed for a Colorbond roof frame, over 600 roofing screws missing, the lack of a treatment plant worth at least $24 000 to install, 44 framing defects including 7 serious ones (where the building surveyor working as an in-house quality control building consultant for the large builder found only the 7 serious ones of the 44 items wrong with the framing; and all of which the certifying building consultant the owner hired did not find), you begin to realise that quite a few if not most building consultants and quite a few if not most registered building practitioners are not equipped to report on new building work at all. A capable building consultant needs far more than just a degree.

But these building consultants (many of them registered building practitioners and architects) were (and probably still are) carrying out inspections for new homes. So what does that say about the residential building industry? The fact that only 4 out of 22 building consultants discovered even a quarter of the defects in the new homes they inspected, makes you realise that building consultants are a large part of the reason why the quality of buildings is so bad today. And (building consultant) opinion report writers are not even regulated. What a system we in Australia have!

So to the people who built these defects into the houses – what about their knowledge / care / lack of supervision? Should they be permitted to continue producing the same abysmal work over and over? Because they are doing just that!

And if Building Consultants lack the knowledge or worse still their desire and/or ability to learn, what does that say about the system that permitted these individuals to enter the field of building consultant (incorrectly called building inspectors generally in Australia)?

Unfortunately many building consultants seem to have no idea that defects can exist when there are no cracks, no rot, no corrosion and no obvious failure.

They do not seem to know that there are far more faults which relate to workmanship, codes and regulations, than those blemishes apparent in materials and finishes.

And these hidden and/or little known workmanship defects are usually far more costly to rectify, often totalling ten times the cost of rectifying visible blemishes.

In my opinion it takes an experienced person 2 - 5 hours to carry out a thorough & considered frame inspection or final inspection, depending on the size and/or condition of the house.

Building Defect

I've heard from several estate agents that most building consultants take far less than 1 hour to inspect a whole house.

I can easily believe that, because I've seen a building consultant take just 17 minutes to inspect a 90 year-old thirty square Canterbury house (and of the 7 building consultants on site at the time, I was the only one with a ladder, overalls or spirit level), so I can imagine that someone looking only for noticeable blemishes may well take a very short time, particularly for a newly painted home.

So we need a definition of defect now. We've needed it for over 25 years... ever since portions of the building industry were privatized.

Chapter 6

Before You Use a Building Consultant, Be Sure to Ask What His /Her Disclaimers Are.

I cannot copy out other building consultant reports without their consent.

But what I can give you is an idea of what disclaimers are commonly contained in many reports of particularly the larger firms, including those you will (or were) almost certainly not told about before you agreed to use their services via their free telephone call.

If for instance the disclaimers say that the consultant will:

- not climb inside the roof but view the roof space from the ceiling hatch
- not climb onto the roof unless it is flat and below 3 metres high
- not enter the sub-floor area of the timber floored house but view the sub-floor area from the sub-floor door

When even the Code for Building Inspections actually states that it is an expectation that the consultant will access those areas if reasonable access is available, I would like to know if those disclaimers are actually lawful. It certainly

Building Defect

seems unethical to me not to tell a buyer about these restrictive disclaimers before you sign them up. I'll leave the legal term for this type of activity to the lawyers.

And what if the disclaimers also say the consultant will not be inspecting metal products and drainage covered under a plumbing warranty, electrical work or landscaping products. Does that mean that metal roofs, flashings, gutters and downpipes will not be inspected!

Does that mean that even though they will get on a lower than 3 metres high roof they are not obliged to look at it? With all of these disclaimers...

THEY WILL ALMOST CERTAINLY MISS THESE VERY IMPORTANT DEFECTS THEN...

This lack of fire – proofing in a party wall...

Building Defect

AND… The cause of this leak…

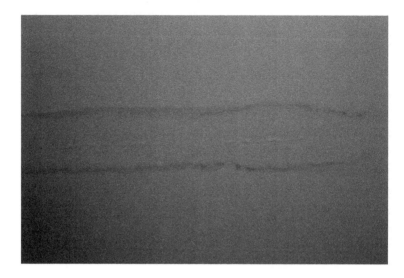

AND these roof leaks in ridge capping apexes.

Building Defect

If building consultants do not get onto the roof or into the roof space or under the house, just how are they going to find potential roof leaks, incomplete insulation, shonky wiring, the likelihood of flooding, impending retaining wall collapse, leaking pipes causing minor flooding under the house, rotten floors under showers or mould under particleboard floors where ventilation is sub-standard? So…

WILL THEY MISS THESE TOO?

This was a 3-year-old unit when photographed and the pointing had already peeled.

OR IF THE LADDER WAS TOO FAR AWAY… will they miss these separated hip end tiles?

Building Defect

I say that a building consultant has a duty to TRY to find glaring faults associated with all of the trades. There could be downlights covered over by insulation or no fire safety barrier under a roof space ducted heater or unearthed power points or shonky wiring.

There should not be disclaimers to cover failings where past employees who didn't try hard enough caused the company to pay out damages for items missed. And disclaimers should not cover for the stinginess of a company not prepared to provide sufficient insurance cover for its employees. Because those companies could franchise out their work to achieve what a single operator can achieve. With all of those disclaimers for whatever reason,

WHAT IF THEY MISSED...

THIS LACK OF INSULATION? Which would result in higher heating and cooling bills for years.

Building Defect

OR THIS CAUSE OF 15MM OUT-OF-LEVEL FLOORS (DIFFERENTIAL SHRINKAGE)?

Whatever the case, do you think you want a consultant who does not look under the timber floor, who does not climb onto the roof and who does not climb into the roof cavity?

******

So I think you'll agree that it is more than reasonable to make this Question Number One for selecting a building consultant to inspect your home…

Question No 1

When obtaining the services of a building consultant, always ask about their disclaimers before you employ them… no matter if the house is a new one or an old one… and do they get onto the roof, get into the roof cavity and get under the house if it's got a timber floor.

There are many more questions that should be asked as well (such as how much do they charge); but this is the main one… and it should be at the front of the code for building inspections in my opinion.

It also pays to employ a building consultant who is experienced in older home inspections, so that your consultant, because they are conversant with the mode of failure of most materials and the time such failure takes, should have an idea on what will go wrong in the future with your new home.

Add the question **how does the consultant define defect**, and you will know if that consultant is worth hiring particularly for new homes or for that matter any home built since 1994, from which date quality began to fall markedly.

With larger firms you may get an inexperienced consultant as well, adding to your woes. However this is not to say that an inexperienced building consultant will necessarily be any worse than an experienced building consultant;

based on the abysmal lack of discovery in 19 of the 20 other consultant reports I have seen and been asked to comment on since 1996.

Basically all a building consultant needs to do is care, think, not be lazy and have sufficient knowledge relating to building products and how they are supposed to be put together. Where a new-comer may be deficient, is in the knowledge of the failure of materials and systems based on experience.

But after the definition of defect becomes common knowledge, there will be a potential problem of old dogs learning new tricks. Perhaps then, picking a new-comer may be the better option.

Whoever you choose, you will need to discuss most of these issues to choose wisely.

The same goes for having houses built for you or buying houses where a series of quality control reports were commissioned by the usually larger builder to give that builder the edge over competitors. My advice is to be wary of such reports which usually have some in-built qualifying statement such as "based on display home quality". This could well limit defects discovered in your prospective home to just those items built worse than the display home, otherwise why do such words appear in those reports? Based on my inspections of past display homes and new ones, both are likely to have many defects. But I say that you should rely on your own capable building consultant with a definition of defect if you wish to know what defects they missed, rather than pay premium dollar for what is being sold to appear considerably better than average.

One of three quality-control consultant reports that I have been asked to check on by the home owners was carried out by a building surveyor who listed just 7 of a total of 44 defects at just after "frame" stage with plastering well underway (better before plastering). I mentioned this earlier in the book, but would like to point out a few of those items he did NOT comment on. See if you think the following defects are worth mentioning...

- Lintels crippled by nails to upper porch
- Non-durable upper deck floor joists
- Inadequate studs to sides of lower storey windows
- 2 upper floor joist spacings 530 mm instead of the require maximum 450 mm
- At least 30 nails missing their mark in just one bathroom floor area (& no extra nails)
- 20mm fall in an upper bedroom
- No framing to the dropped ceilings at all... (Unbelievably) the plasterers simply pre-formed the plasterboard into right angled pieces and siliconed the plasterboard to ceiling and wall to form the dropped ceilings. They were literally waiting to fall off!

I said this inspection was 'frame' stage, but the garage was not even commenced... because the builder wanted payment earlier than he should have.

Another home I inspected after completion, where a larger firm of building consultants was employed to inspect that

Building Defect

home at each stage, still had all of the usual commonplace recurrent defects that I come across in most houses. It was a little above average quality, (but not by much... and of course the framing and slab could not be inspected adequately). I have however seen some good quality houses built by a small number of large builders, but generally volume-builder-built homes are below average... and average is not good at all. (I think that such volume builder built houses probably had an experienced builder (unlimited) supervising the construction, which is quite rare nowadays).

It's just seems common sense to obtain the services of your own building consultant (having made sure that he/she is a capable one), to carry out these inspections on your home, rather than relying on the findings of someone paid by the builder.

When having your house built, stage-inspections (inspections before you pay at completion of each stage) are very strongly recommended, particularly at frame stage, after which framing will be mostly covered up.

If your house is being built by a builder who is also the carpenter or the plumber, you will almost certainly get quite a bit of experienced quality supervision. But if you are one of the 90% who do not have that type of builder, I strongly suggest you get a building consultant who defines defect and ticks all of the other boxes in Chapter 8.

Because, what your consultant misses as regards reasonable life expectancy of all of the components of your house, will mostly end up coming out of your pocket after the warranty period ends.

To those of you buying a home completed after about 1997 (possibly as early as 1994), much of the information and advice contained in this book applies to you too.

To those of you selling after the warranty period ends, there may be less value in your home than you think (or have been advised), particularly if you have started to fork out big dollars for maintenance.

Regular maintenance should not be necessary on a house less than 7 years old.

It generally only happens to houses that are lemons or those that contain a considerable number of components that were short-cut by some of the tradesmen, often band-aided during the warranty period by the tradesmen sent back by the builders at no cost to the builders.

This is an admission that the builders knew that the tradesmen had short-cut the work, and that they knew that it would occur again, after the warranty period had ended.

This heavy reliance on maintenance is discussed in the next chapter under the heading 'time bomb', but there are worse defects to consider than just the short-changed ones.

Chapter 7

Time Bomb. Many Buildings Will Fail Prematurely After the Building Warranty Ends.

When you inspect new houses you must consider the future. Usually there are a number of possible events that have not yet occurred. The houses have therefore not been fully tested... and that is vital for all building consultants to understand.

Failure of houses can be caused by one or more of quite a number of factors.

If you take a walk around an estate just a few years old (as I did once); it soon becomes obvious from the footpath, that quite a few of the houses have failing elements, even during their warranty periods. None of these elements have lasted what is considered to be a reasonable lifetime. The following were noted more than once...

Firstly the rot sets in to any piece of exposed non-durable timber including exposed plywood garage door lintels.

Some house walls are built on the boundary and constructed of just painted (and hopefully correctly sealed) fibre cement board sheeting over 2 layers of plasterboard with foil between... all on a timber frame, with no access.

How long is this type of wall supposed to last I wonder? Because excessive footing movements can open up the protective cladding and expose raw plasterboard! How good will the required fire rating be then? And was this system thought up without regard for reasonable life expectancy, just to keep the price of houses down?

Turned verandah posts (usually oregon because it splinters less when turned), are invariably not oil-base primed (particularly to end-grain and around the bolts or coach screws) and quite a few will therefore have a life of somewhere between 4 and 15 years when they can actually become quite unsafe.

Roof tile final pointing coats are peeling off some of the ridge tiles and the base packing is starting to erode... and the hip tiles on a few have separated, often before the peeling has started. (Occasionally hip and ridge tiles are whipped up in a wind gust sufficient to make them become airborne... but only to a few house roofs). So should we just ignore this potentially life-threatening premature failure?

The part-tiles over the gable end soaker flashings often slip out (the owners often being unaware). The water runs down the gable end eaves lining and causes staining and possibly rot if timber fascias were used. Now with so many houses having no eaves at all, this slipping part-tile causes dripping onto the brick skins, which sometimes runs into the cavities of the brick veneer walls and commences to rot wall framing or even plasterboard inside those houses. (I've seen such rot occur in a vital position to studs under a beam support).

Paving has been placed adjacent the external walls sometimes over weepholes and the house is now at risk of termite attack, even with its deterrent spray under the slab.

This is just a sample of items that are visible to the trained eye from the footpath.

******

These next examples are frequently encountered; and they make you wonder just what else could go wrong with houses after the warranty periods end.

Timber window sills lifted by the brick sill pressure, (caused by the lack of consideration for shrinkage of the unseasoned hardwood floor framing that was used), hold water sufficient to keep the timber sill beading moist for long periods of time (the gaps at the beading are not sealed by the acrylic paint system). Sometimes the windows start to rot... in just 5(+) years... not helped at all by the poor paint job. Many are substantially rotted in less than 10 years, (but the warranty is well expired by then).

Garage plasterboard ceilings often have a considerable number of popped nails and the cornices can also be adrift to nearly half the perimeter. The fact that:-

- many garages had slabs poured after the walls were built,
- the walls are in effect just a single brick skin with joints of that porous stuff called mortar soaking up the incident rain,

- there is actually no requirement for a damp-course at the base of garage brick walls,
- there was no underlay placed under the concrete in-fill slabs,
- there is no insulation above the plasterboard to protect the joints,
- and the ventilation openings to a garage are often 10(+) times that of any bedroom...

(to me) means that the garage plasterboard ceilings are going to be considerably moister than the plasterboard ceilings in the houses attached to those garages, including those in wet areas. The reason they are plasterboard ceilings is basically fashion... its faster to install and neater than a fibre cement ceiling. But plasterboard ceilings in such situations are basically not fit for purpose and are therefore defects.

Soil reports often contain precautionary advice against the planting of trees close to houses, and most (but not all) contain recommendations regarding trees recently removed. These recommendations are sometimes ignored by building surveyors and builders (even when the soil is classified as P (meaning problem)... requiring the footings or slab to be designed by a structural engineer. But when these classifications are ignored or misinterpreted, or designed for the perceived problem and then ignored by the builder who knows better (and the certifying building inspector for one of several possible reasons): then the houses are at risk of future significant soil movements, based on many houses I have come across.

Building Defect

A row of cypress pine wind-break trees could be removed or die a few years before a housing sub-division is proposed flanking such a past row of trees. And if drought conditions prevailed during this time, as would have been the case for much of the past 23 years, then soil engineers may not realise that this row of trees ever existed. Sometimes such past rows of trees can be as little as 10 metres from where units or sometimes even houses are to be built… close enough to cause significant future damage.

The same would apply to trees removed during a drought for unit sites, to large trees retained by order of councils, and by the rapid growth of trees recently planted nearby.

Some the house slabs near such a past row of trees or large removed tree(s) can lift and cause a marked fall in the house slab, away from the tree(s) particularly after the drought ends. And those house slabs near large retained tree(s) can fall markedly towards those retained tree(s) if the drought continues. And sometimes these movements occur years after the warranty period ends.

The situation can be even worse for houses with strip footings.

Houses in these situations can virtually split in two. Large 'Zorro' type cracks open up in the plastered walls and diagonal cracks up to 10 mm appear in the brickwork... some overnight. So the warranty period can be insufficient time to fully test these house footings and/or slabs... a great shame considering that these elements are the most vital part of the structure of those houses. This ignorant construction is grossly defective; and it is the duty of every building consultant to attempt to discover such defects... preferably before they occur. They are very difficult to diagnose particularly without all the relevant documentation concerning the footings. And it would be so much better if these problems could be averted before they are even designed and built. Many times there is very little that can be done to help.

Trees dry out soil sometimes down to a depth of 2 metres... possibly more. The soil nearest the tree is bone dry and the further away you go the less dry the soil becomes. During a drought the soil will dry once more and shrink still further. And when that drought ends, the soil slowly regains its moisture and will again swell... pushing up footings in the affected zone to varying degrees.

Soil tests in recent times may have under-stated the degree of possible movement in soils on some recent developments to the north and west of Melbourne: and floating slabs such as waffle-pod slabs designed for those under-stated criteria can as a result prove be grossly inadequate.

Bathrooms often have leaking showers, (another story detailed elsewhere in the book because it is such a common problem). These leaks allow water into the unsealed MDF skirtings and architraves plus into the particleboard flooring. All of these swell fairly rapidly and are accompanied by various grey and black moulds. As the particleboard floor sheeting swells it pushes up and tilts floor tiles. As the leaks continue in the wall joints, this constant excessive moisture loosens wall tiles, rots plasterboard, and then commences on the framing to floor

Building Defect

and walls, eventually dripping onto the ceiling below and rotting that. I have seen this scenario over a dozen times during the warranty period. (Sometimes water runs down the walls below and starts rotting the carpet).

But what if the symptoms of failure show up too late for the owners to make a claim? Surely then the building consultant has the responsibility to diagnose this likely future failure based on inadequate workmanship in the bathroom wall tiling.

Well I seldom see any mention of this in other building consultant reports… because they do not have an adequate definition of defect. Also there is a 10 year negligence warranty by all who allowed the system to be installed incorrectly. Beware all people concerned.

Once, a small boy and his friends were hanging onto a second storey balcony when the bottom rail of the balcony came adrift and they (all 3) swung out. Luckily the top rail end nails held. The owner nearly had a heart attack. It's just as well he was a very fit person and temporarily stabilized the rail.

These examples, the above story and the list of new products and systems discussed in this chapter are here to show you that there is great need for questioning minds to appraise the suitability of most house components. But it just isn't happening.

******

Now I'll discuss some recently approved materials already commonplace in the industry.

L.O.S.P. TREATED PINE

LOSP treated pine (a fairly recent product steadily replacing CCA treated pine) is starting to split, not just at nails but sometimes at edges and on the top surface. Where it hasn't been protected at cuts (in considerably more than 90% of cases is my guess), core rot can occur… and can be quite rapid I believe… and together with the splitting, could perhaps become a significant cause of structural unsoundness in the not too distant future. (We'll have to wait and see). I personally have had occasion to see an L.O.S.P. stair tread split under foot.

Now this product will probably last a long time when installed in accordance with the manufacturer's recommendations. (So that must surely make those recommendations requirements, right?).

But how long will the material last when installed in an unworkmanlike manner, as it so often is… with no end treatment? And this timber is quite often used on timber balconies 3 metres off the ground, and people trust it to last virtually for ever. Some of you possibly agreed to the use of this product, knowing very little about it… but the worst thing about this issue is that very few builders and building consultants have bothered to enquire about its longevity even when installed correctly. Most installers have probably not even bothered to find out about the recommended (required?) treatment of end grain… and read the small print of the specification… and boy is it small print… and it's white writing on medium blue paper! Osmose H3 is the product I believe; marketed here in Australia under the trade name Protim. It has been used in

Building Defect

England and here for years – but perhaps the end-grain is treated there... every saw-cut. I don't think it happens much at all here in Australia though. (And that's not all either.)

Splits in deck joists will develop core rot.

Building consultants may think that simply noting that a durable deck framing product has been used is sufficient examination... well it certainly is not... there still has to be suitable top-edge protection over the tops of the joists... one that seals around the nails for a considerable time to protect against the likely splitting effect of penetration of the top edge by nails... and through the protective 5mm outer portion of the structural joist members in to the ordinary un-treated radiata pine core. And everyone calls this product treated pine. Well it's not at all like the CCA treated pine which is treated right though the timber.

Building Defect

Use LOSP treated pine in the wrong place and without end treatment and it will rot.

The product is good... but usually the workmanship is poor... due to ignorance most likely... thanks in no small part to the poor definition of specification in virtually all building contracts.

HEAVE AND ARTICULATION JOINTS

Perhaps builders, building surveyors (called relevant building surveyors in Australia for some unknown reason), building inspectors, building consultants, structural engineers and many soil engineers think than they have more expertise regarding 'heave' than the writers of TN61 and The (Australian) National Construction Code once The Building Code of Australia (BCA). To me it certainly appears that way. Heave and its possible ramifications must be considered it says in the BCA. And because of the long-term recent drought in the soil, it means that many footings of houses built since 1995 have yet to be fully tested because the moisture takes time to fully return. Heave, although to date a less common problem than subsidence, nevertheless occurs every time cracks in brickwork begin to close back up. And it is very difficult to prove that soil deep down will not heave unevenly one day, particularly if the history of the block of land (trees, drainage, past buildings, rows of trees and the presence of springs), and layers in the soil is not known. Therefore the expertise that results in articulation joints being installed incorrectly in virtually all houses built since that date cannot rely on much experience (if any at all) as regards possible future heave.

Articulation joints are there to confine settlement and heave movements (within designed limits) to these joints: in the case of heave, to avoid considerable compressive movements on the windows or adjacent masonry.

The likelihood that this condition will occur for tens of thousands of homes, resulting in serious damage to some of those homes, seems to have been swept aside by the permitted (so-called) *expert judgement* of builders, architects, building surveyors, soil engineers, structural engineers, designers who permitted this commonplace short-cut to take place: as well as by the building consultants who do not list these wrongly constructed articulation joints as defects.

In the name of keeping prices down, we are now left with a legacy of potential major damage (much of it structural) to quite a considerable number of houses. The drought which just ended a few years ago (in 2010 in Victoria, Australia), exceeded any previous recorded drought by a long shot. But it didn't last forever...so builders and registered building practitioners should not have relied on the myth that every window creates a full height articulation joint, because that only covers single settlement at corners... only half of the total equation... and has resulted in a defect in nearly every home built since it began in Victoria at least. I wonder if anyone other than me (including structural engineers) has ever thought about this issue at all.

The commonplace lack of full articulation of brickwork, that I have a duty to point out on virtually every new house I inspect, will be glossed over (denied by default) by the courts, simply because no one has yet shown, in a VCAT hearing, that the lack of such full articulation was a problem for a particular house, even though it is likely to get to a VCAT hearing very soon... and even though the Building Code of Australia clearly spells out how that articulation is to be installed and that heave MUST be considered. This is another instance of where expert judgement (clearly not expert at all) is permitted to rule... Just as it did and does for apartment towers and office towers throughout the world as regards at least flammable cladding.

Up until now, virtually all of the registered building practitioners (Australian terminology) have ignored this wording.

I have personally seen 4 instances of substantial heave; and the damage done was far worse than for the same (substantial) degree of settlement. I have one such case on the go at present and the opposing structural engineer experts have virtually admitted that heave of at least 50mm has occurred, but do not seem to be in agreement as to the cure. That is because there is no perfect cure and the most appropriate is most expensive.

What a joke the industry is becoming!

The industry is continually permitting that which should not be given a permit. And it shows clearly to me that defective work is not really understood at all by authorities, the barristers, the solicitors, most building consultants, most soil engineers, most structural engineers, most building surveyors, most building inspectors, most relevant building surveyors, most builders and/or their bricklayers... because this watered-down version of articulated masonry is being wrongly installed in virtually every house right now, and has been since at least 1997, with the apparent approval of all concerned (some by default).

It also clearly points out that 'defect' should embrace the concept of possible failure when it cannot be proven that the failure will NOT happen, by the transgressing so-called 'experts' who installed it or allowed it (via their judgement) to be installed, contrary to the expertise contained in the relevant codes or technical documents. The BCA frequently used words 'shall' & 'must...' and really need to be heeded.

I'm not a QC, but I think it is about time that someone out there challenged the widely held belief in legal circles that the onus of proof is always on the claimant, not the respondent... based on the reasoning under performance provisions in the National Construction code previously the Building Code of Australia.

So Brave Client Barrister, there's a challenge for you – if you can understand what I've written in the previous few paragraphs. Help me challenge this concept, for the good of the industry and the good of houses... and make a name for yourself in so doing this (I say) vital work. Otherwise there may as well not be any Standards or Building Code at all... and in the not too distant future the way things are going, that may just happen. I think the authorities may already have started to differentiate between relevant codes and optional codes... what next?.. 'houses as specified' (but nothing on workmanship contained in such specifications)?

******

This next example (not a new house, but a house built in the early 70's and viewed by me in the late 80's during a year-long drought), is one that took me 7 hours to figure out.

As I said, this house was not a new one; but it seems that what happened to this older home could just as easily happen to (possibly) thousands of new homes, (and indeed has started to happen to 3 newer homes I inspected recently). It represents what may be the single worst problem a house could encounter (apart from earthquakes, floods and hurricanes); namely 'HEAVE'. (And this older home story is better than the new home stories because the builders did take some action, although inadequate, to prevent settlement).

So to the older house story....

******

When I arrived at the site the owner showed me the undulating interior of his brick veneered house that had recently been under-pinned and repaired because of... no kidding... 20mm width cracks. These are sizeable cracks and fortunately fairly rare in outer eastern Melbourne, Victoria, Australia (Bayswater) 'M' class soil areas.

And in just 9 months since that under-pinning of the footings at each end of the north wall, the brickwork had cracked again... wait for it... 20mm again... in two different locations.

The only thing that the owner had done, (I discovered when questioning him about recent works), was to cut down 2 of 17 boundary line cypress pine trees to let in a bit of light to the kitchen... twelve months before it first cracked and was subsequently underpinned. The trees were just 4 metres from the house at the Kitchen wall.

I had never seen the results of simple heave in the absence of settlement at this stage of my career, but was almost certain that the cracking must have been soil moisture related. I reasoned that the soil moisture was returning to the soil opposite where the trees had been removed... because the ends of the north wall were already underpinned (and therefore not likely to move much at all from then on). Amazingly the brickwork joints were now out of level by about 38mm. The moisture was not being absorbed out of the soil as is usually the case when there are large trees near cracks in brickwork.

Whilst walking around the house, I discovered that there were virtually no other cracks in the house brickwork. The problem for the brickwork was confined (it seemed) to the north side only. It must have been either the settling of most of the house or the lifting of the central third of the north external footings... and walls... and roof... and the middle of the house near the lifting walls... with the resultant failures of linings, doors and joinery inside.

Removal of the trees was therefore the likely cause. After considerable thought it became apparent that the trees over the years had each obtained their water from a narrow strip of land each side of the row of trees. The CSIRO (once used as a preventative research station, but now used to promote new industries) published a booklet (10-91), which mentioned rows of trees as being more of a problem than a single tree: and I therefore reasoned that the tree roots of a row of trees must have to travel considerably further to obtain their water because they could not compete tree-to-tree. (You might not be surprised that this information contained in the CSIRO document has been superseded by a series of Technical documents which lack some vital components of this information... how about that for dumbing down the public... and future experts).

And the revelation that the trees were planted long before the house was built concentrated my thinking on time-related issues. The trees had historically dried out the soil deep down prior to the building of the house. Now this moisture had returned.

Building Defect

Why then were the eaves on the South (far) side of the house tilted at more than 5 degrees whilst those on the north side were still horizontal? Now I think that most consultants would not have noticed the tilted eaves on the far side, because there were no cracks and no obvious failure.

It wasn't until I had made a model out of staple jointed balsa wood that the tilt made sense. The forces acting on the north (cracked) side were basically (as Spike Milligan would say) "in the direction of up". But the forces on the other side were compounded – the push was 40mm at the north wall, but also 20mm in the middle of the house under the ridge and 30mm under the north under-purlin props, but there was a great resistance at the south wall joint because of gravity and the ceiling joists tying the frame together. Hence a simple joint rotation-rafter-tilt was the result. The skew nail joints had twisted and permitted a simple tilt of the eaves on the south side.

And it then dawned on me that the under-pinners had under-pinned the portions of the north wall footings that had NOT moved!

There was no cheap cure because the trees were there long before the house and the original builder had given no thought to soil being historically too dry due to tree root influence.

And the under-pinners had wrongly assumed that settlement of the ends had occurred. Complete under-pinning was now the only cure, apart from re-building of course.

Heave is quite a frightening force. You might have seen a mushroom push through bitumen. Well heave does the same to buildings. And the cure is so expensive and so difficult to achieve in a workmanlike manner... because even when the remedial work is done, how can one be certain that the entire heave has occurred? That is why it is so important to safeguard houses against such a possibility. That is one of the reasons why soil reports are carried out...

But I often see soil reports that seem to gloss over this issue... **and simply recommend actions instead of insisting on those actions**.

******

This above example, the newer homes with heave and the drought are strong reasons why I think that failure of footings and slabs could become a huge issue facing home owners, building practitioners and the whole residential building industry in the not too distant future.

And If it occurs, let's hope that it does not fall under the pass-the-buck industry mantle 'matters beyond our control'... even though it is well within our control. Owners may well be asked to put up with the large loss in value... all in the name of lower building prices & progress. Governments may even legislate to remove home owners' rights in this matter... I do I hope that never occurs, because owners' rights are almost non-existent as it stands... and...

Building Defect

Because Australia's National Construction Code (once called The Building Code of Australia has always stated that heave MUST be considered.

ROOF TILES

Tiles over soaker flashings often slip and cause leaks onto the blown-in insulation, but will possibly not be noticed until after the warranty period, because the insulation absorbs the drips, particularly Insulfluff. So it may well be argued, (wrongly), that the system is performing OK, and is therefore fit for purpose. The persistent dripping will eventually get through the insulation and damage the plasterboard ceiling... *old man time* again... remember I mentioned the lack of consideration of time by most building consultants.

Also, the soaker flashings are often kinked (due to insufficient strength), after tiles have slipped out above them and have been pushed back into place. These often leak noticeably.

Flexible pointing is often applied too thinly – nobody knows much about it – so it peels and separates and fails to hold ridge tile to ridge tile (because often they didn't bother doing the collets at all): and this totally negates why it was permitted to replace the BCA requirement to hold down the end 4 ridge tiles on roof ridges in the first place. SO, now they can blow off – millions of ridge tiles state-wide! And the top fixing rows under these ridge tiles are usually not fixed either.

Building Defect

This flexible pointing is just over one year old and paper thin at the tiles instead of the required 3mm - 4mm thickness. There's more wrong that that too... maybe later.

The next photo shows the lack of fixing of the typical top fixing row of tiles with a highlighted tile 2 rows down. The top rows are also fixing rows but have not one fixing.

Building Defect

LEAKING SHOWERS

I've earlier described in detail how leaking showers are on the increase. And a lot more of them are going to fail after the warranty period is ended: some having not been used until the young couples decided to have some children.

Plasterboard (less durable than fibre cement board), usually without corner flashings, when combined with particleboard floors (far less durable than floorboards), should mean that we need a better system to cope with the possibility of leaking showers; you would think.

Building Defect

This case was just 4 years old so it doesn't say much for the durability of MDF. So surely it's incorrect to permit its use in wet areas and possibly garages too without priming all edges?

Some showers don't leak for a while, because silicone even when incorrectly installed across the junction joints usually takes a while to fail. Others may also leak when droughts end or soon after they start, due to structural movement when the foundation soil (under footings) shrinks and swells differentially.

But the point this book is trying to make is... that showers should not leak; they should also have been prevented from being able to leak... and every leaking shower has been installed by tradesmen (in a tradesmanlike manner as distinct from in a workmanlike manner), supervised and approved.

Building Defect

TIE-DOWNS FOR METAL ROOFS

Some roofs have lifted off in fairly recent times in Victoria (Australia) – having heard of at least 3 during the last 15 years or so – a Brighton garage roof, a north eastern suburb house roof and more recently a Malmsbury house roof. So it is important, for safety reasons alone, that tie-downs be adequate for the fiercer more frequent storms that Melbourne now experiences.

An example of inadequate tie-downs follows this story.

Why do you suppose that the builder's original building consultant, the building inspector and the building authority-appointed building consultant who replaced the original building consultant all said that these pictured tie-downs were adequate in their opinion when they did not comply with the Timber Framing Code AS 1684, even when based on the low design wind speed very likely considerably lower what it actually was?

Because they had no definition of defect I say... and they therefore had scant regard for time.

There was a chance that the inadequacy may one day affect the house roof and possibly even the safety of the public. What they had in mind, it seems to me, was to remove the large cost item being claimed to rectify the defect, because there was no evidence of movement up to the time of the inspection. But the house was just 4 years old. This was to me a sort of cronyism at work.

And I say that the system that allows these people to continue acting in those capacities is grossly inadequate.

Building Defect

And yes it was a large building firm who skimped on the tie-downs, using a supervisor who was not an unlimited responsibility type builder.

But there was also a worrying misunderstanding of how winds can affect roofs by at least the building inspector and the supposedly well-credentialed authority appointee building consultant. (A description follows the photo).

This a ceiling joist hanger and it is not approved as a roof clip, but just as importantly it only ties the rafter to the top plates and the tie-down is required to be fixed to a stud with at least 3 well-spaced nails. The photographs taken by the owners at frame stage also showed that no plates were installed tying the studs to the top plates as is usually done for factory-made wall frames. And the negating authority-appointed building consultant had copies of this and the frame stage photos. And yet he denied the defect for the basic reason that there had been no apparent movement!

Well freak (20 / 50-year return) wind gusts come when they come, but rarely by the time a house is 4 years old for obvious reasons.

With a proper definition of defect made public, this fiasco would not have occurred.

******

RECENT PRODUCTS GENERALLY

So if any of you readers are thinking of using recent products or systems such as acrylic renders, slab lifters, post construction water repellents, membranes for wet areas and tiled decks, 'the latest' termite deterrent spray(s), the lack of sill flashings to aluminium windows, timber protection, polystyrene cladding and so many more... read the warranty conditions and think before using them at all, because some recent products and systems have definitely not passed the test of time at least until the effects of the drought are well and truly over. (Under-tile membranes, for instance, do have failures by admission of their manufacturer's own installers... and those failures occurred during the recent prolonged drought conditions where structures had not been pulled around and distorted anywhere near as much as they might be after the drought ended or when another one started). And what if there is no warranty or code that covers the use of such materials?

Some aluminium windows installed with wind flaps and alongside wall sarking, have failed miserably on several weatherboard houses where there was no cavity to help out (as is the case with brick veneer houses). But that doesn't necessarily mean that they don't leak into the brick veneer cavities sufficiently to cause problems in the longer term... for instance the cavities may well be bridged here and there by mortar or service pipes or electrical wires, which could divert the water to the sarked wall frames and to places through the sarking onto wall framing or worse still, the edges of particleboard flooring under wall frames.

Builders, designers and building consultants are all required to find out if and how these products and systems are inadequate (for the particular use in mind), when there are short-cuts commonly taken and when there has been ignoring of the manufacturers' additional requirements and/or recommendations. Otherwise we should perhaps not use them or permit their use.

These materials could all do with a lot more considered thought.

FLASHINGS

Flashings are installed to weatherproof joints. They usually have to allow for movement. And movement is caused by several factors including wind, shrinkage of timber framing, stump rot, temperature changes and sometimes soil conditions under footings.

So it stands to reason that if a cover flashing is firmly secured to roof sheeting, and silicone sealed to the brickwork (sometimes even if fixed); then when movements occur, something has to give. Now silicone can withstand just so much tension before it tears... as it does... regularly. (Silicone is also not approved by the

Building Defect

manufacturer as a seal against mortar but is regularly used that way).

Roof plumbers, in order to cut out time-consuming work, used to install the silicone sealed cover flashings over the lower flashing upstands, but now often install just the upstand flashing silicone sealed to the brick, (or in the instance below under a gable roof, inadequately sealed to the fibre cement wall cladding, with no cover flashing at all. The silicone in any such installations fails rapidly.

I explained earlier, that the Plumbing Commission permitted an easier solution for roofs meeting brickwork situations, but the roof plumbers have chosen to skimp on the greatly reduced labour component still further, making many of the new style joints fail or likely to fail prematurely. This short-cutting is not expert judgement; it is exactly the opposite and is not permitted under the alternative solution requirements of the Building Code of Australia.

******

And so, the term commonplace is warped to the word standard... and workmanlike is blurred to seem to be whatever workmen normally do... all in the name of progress & cheaper housing.

But workmanship demands adequate knowledge and adequate considered thought in order to avoid the pitfalls that can occur. Tried and tested is a good rule when in doubt.

There are many more examples of potentially expensive-to-rectify impending problems that I could list, but I feel that enough has been said to make you realise the importance of careful labour... and that...

The detailing of the labour required to achieve good workmanship needs to be included in specifications.

It is very, very important and should not have been virtually discarded as it has been these last 15 or more years; causing the dumbing-down of the industry right to the top it seems, helping create a blight in the residential building industry of possibly huge proportions.

Relatively maintenance-free housing seems to be fast becoming a thing of the past. Many of the larger builders are maintaining houses several times during the warranty period (instead of rectifying the defect). The likelihood that many of the time-based possibilities have not been adequately considered by the building authorities or by the building practitioners in charge of designing, checking and building your houses is of great concern to me.

Building Defect

And if something is not done about future problems very, very soon…

        The time bomb is NOT going to be large,

                And it's NOT going to be huge,

                        It's going to be ginormous.

Chapter 8

Greed and Privatization of the Building Industry Have Caused Considerable Blight

There have been some massive changes in the Building Industry, many of them not advantageous to the long-term economy of Australia.

The *turning of a blind eye* by certifying building inspectors, so as not to upset the builder / (relevant) building surveyor relationship (and therefore in turn the certifying building inspector's livelihood), sometimes saved the builder a fair bit of money. It takes just a little friendly pressure at the right time. And I think that this was occurring a great deal, (or just as bad, many certifying building inspectors have been downright inept).

But if I'm correct and this did occur, then this was a form of cronyism... privatized cronyism.

I've attempted to argue against this scenario in my mind, but could not explain why the installation of insulation in nearly every roof space is incomplete and non-compliant, why nearly all articulation joints are incorrect in at least one way, why the wall tiles in wet areas are incorrectly jointed, and why nearly every roof tile installation has inadequately fixed tiles and inadequate pointing.

Building Defect

In about 2016, the Victorian Government forbade builders from choosing their (relevant) building surveyors, making the owners hire them instead. (Other changes made at this time will be discussed later).

Many builders are not retaining or draining cuts on boundaries and steep embankments under houses: but some certifying building inspectors are approving this type of work, (work not done in compliance with established civil engineering principles), supposedly in accordance with their so-called expert judgement: permitted (they think) under the National Construction Code (formerly the Building Code of Australia)... but really not permitted at all in most instances where this occurs. It is possible that the certifying building inspectors do not even look at the job long enough to find all of the defects, but how could so many of them miss over $100 000 worth of defects that occur in a frighteningly high number of houses... or over $50 000 worth of defects in additions jobs. Because I've found this situation in close to 8% of the new houses and new additions jobs that I've inspected. That's one job in eight! And even if this is a skewed result they still happened.

Builders often offer to obtain their own experts to assess sizeable wall cracking in their recently built new additions for instance, and these experts invariably fail to assess the causes adequately: and instead opt for monitoring & maintaining the cracks, but not curing the cause of those cracks.

Several times now, I have had to battle with this type of builder's expert approach, where falls in floor levels exceeded 50mm in 'M' classified sites; and still no solution had been proffered by the builder's experts after considerable passages of time often exceeding 18 months. Mostly the cause of the grossly excessive out-of-level floors had been incorrectly diagnosed or grossly underplayed by structural engineers who called themselves foundation experts. Expert negators they may be, but by ignoring vital parts of the Code for Footings (and using just Appendix C), they were nothing short of negligent in my opinion. And yet they continue being permitted to negate… because the owners did not proceed to a VCAT Hearing, but instead took this easier route, eventually accepting about half of what it would take to cure the problem… and of course, accepting a gag as part of the settlement.

So cronyism is alive and well… thanks to our legal system that protects enterprise.

People in the industry may worm their way in to being recommended by the authorities, but it does not mean that they are experts or fair-minded people. You must ask them pertinent questions to make sure before you employ them. And I suggest you take a serious look at the pertinent sections in this book first.

The ex-Building Commission was replaced not long before this book was written. But I don't think it was because of the Building Commission's failure to curb the drop in the standard of houses being built. So this book remains vital despite this change in its leading authority.

Privatisation in 1990 (1994 officially) of the building surveyors with the new title of relevant building

surveyor (and certifying building inspectors working for those relevant building surveyors) has not worked at all as regards quality of product. Quality has declined rapidly and remained at a very low standard since about 2004. Relaxing of the standards has also not helped at all.

And building warranty insurance was also privatised in about 1994 soon after building surveyors were, with the Housing Guarantee Fund Limited being phased out until all warranty periods had ended.

People in the industry, (including quite a few builders, tradespeople, certifying building inspectors and unfortunately many building consultants), began to accept what they saw every day as normal and inevitable; and so poor workmanship actually morphed into the acceptable norm in many peoples' minds.

Negating consultants are actually using the word 'standard' incorrectly to defend pure 'Band-Aids'.

So it is vital to understand the difference between *'commonplace'* and *'standard'* plus the difference between *'tradesmanlike'* and *'workmanlike'.*

It will help considerably to stop the lies perpetuating in the building industry.

Remember, all of the Governments of Australia, federal and state, are relying heavily on the Building Industry to keep the economy buoyant (even with a depressed rural sector). Why do you think that there are so many immigrants? It's probably because they all need housing, and will keep the economy robust and the GDP close to 4% as per Keynesian theory... they undoubtedly think.

So it would be a 'career-threatening-move' of any government to upset the boom times. In fact they seem hell-bent on freeing up the industry more. That may be why the Victorian Government removed the word 'control' from what was once called the Building Control Commission.

But this gave the wrong impression to those who wished to concentrate on the holy dollar.

Freeing up the industry is thought to permit more innovation and helps to restrain prices. Well it certainly seems to have achieved the innovation part... but not in the way expected. Some restraint in pricing certainly seems to have occurred considering the larger demand during the recent boom times. De-regulation also helped to keep prices down, but at quite a cost!

Based on a growing number of cases that I have been involved in; where my clients have come up against a series of obstacles which seem generally to prevent home owners from getting to a court hearing in VCAT, I believe that the system in place actually obstructs home owners more than it protects them. That is what so many others in the industry feel too. The system tends to protect enterprise often at the expense of fairness to home owners. And somehow the public outcry that should have occurred has not.

VCAT is directed by the Building Act 1993 to place the emphasis on mediations rather than simply having a Court Hearing for cases where the claim was over $10 000 (recently raised to $15,000, but should have become $20 000 some time ago to keep pace with costs of building... and should really be double this amount to be a fair maximum). I believe this to be our system. And if mediations were fair, then I can see the point of this legislation.

More recently (I think to reduce the large expenditure by both new home owners and builders in full-blown VCAT disputes) in Victoria (Australia) at least, another 2016 reform was instituted as a compulsory first-stop dispute forum called the Domestic building Dispute Resolution Victoria (DBDRV). If anything this has merely further repressed new home owners still further and has as its publicised main aim 'the expediting of each and every dispute.' Basically this DBDRV is there simply to make the disputes go away... and the conciliators in charge do not even have a definition of defect, relying instead on the head-bashing (builders and owners) and (costs) frightener techniques used in VCAT Compulsory Conferences for its results.

And it works to a large extent, but to the detriment of the defective houses, because the (gagged) agreements reached invariably result in the band-aiding by the builders of the substantially pruned list of claimed defective works... to the point that the builders can send back the tradesmen who originally short-cut the works to short-cut the works all over again (virtually free of charge to the builder and again poorly supervised)... leaving the cause of those agreed defects to fail prematurely once again in the not too distant future.

But I say that both DBDRV and VCAT mediations are not fair at all... and compulsory conferences are even less fair. Mediators, before the Mediations commence, openly state that you cannot define defect. So in order to resolve disputes, they resort to pressuring both sides, because nobody can agree what a defect actually is. And home owners tend to give up sooner than builders.

The VCAT (Victorian Civil and Administrative Tribunal) cases I was involved with and that eventually went to a hearing, had 1, 2 or 3 Mediations, a Compulsory Conference and a Directions Hearing. (I know of one case where the solicitor wanted to handle the case alone had 9 mediations at great expense to the owners. He had no definition of defect).

Either an alleged defect is a defect or is not a defect. It's pretty simple. That has been and is still the main stumbling block in nearly all disputes... which is why I have provided that definition.

Some hearings last less than a week, some last for a few weeks and cost a small fortune. But the compulsory mediations and compulsory conference also cost a lot before you can get to a hearing.

With the definition, hearings would not need to take anywhere near this amount of time if the definition of defect removes the possibility of argument about whether or not an item is a defect. Basically there would only be the need to sort out the costs to rectify the defects on the basis of a reasonable scope of works. It would take far less court time, barrister time, and expert-witness time to achieve a fairer result.

So there could be a strong argument for dispensing with all but one of the compulsory meetings for instance, one basically to confirm the status of the alleged defects, which would also bring the parties much closer together, requiring just fair scoping of works.

Cases under $15 000 (where there is no definition of defect) are I think over in a day. Each party pays their own costs. This figure could perhaps be increased to say $30 000 or more, and still be over in a day with a definition of defect in place. This should drastically reduce the costs associated with every dispute, and cover almost all disputes except those about houses built so badly that their appropriate description would be 'lemons' or 'doozies'.

One big difference in mediations versus court hearings seems to me to be that the name of the builder, which could be harmed as the result of publicity associated with a VCAT hearing, is far less likely to be harmed in a mediation result, because of the gags permitted by the system in the terms of settlement. The agreed amount is then not permitted to become public knowledge and home owners are probably advised not to tell their story. In mediations, nothing has been proved, and so the easier route of settling in mediation process results in the protection of enterprise. (And this system could still exist if people and business want it, but under my proposed system, the only arguing left for on-going mediation would be how much to rectify the agreed defects to a reasonable standard.

So the mediation process is usually good at least for business it seems.

Now the whole idea of the mediation process is to make the parties come to terms without going to a full hearing in VCAT. (This could be invaluable in cases where the experts on each side are inept and/or either participant is extremely unreasonable and can be persuaded to be reasonable). But the mediation process still costs a lot and means compromising again and again… and again, until everyone is relieved but basically unhappy. And the settlement agreed to is far less than it costs to rectify the defects properly. So houses are invariably left with many defects left untouched or merely maintained, so that repeating and frequent maintenance will be required from then on.

Less public knowledge due to the required secrecy of the VCAT compulsory conferences means that less builders are thought to be building poor quality houses. And less disputes getting to a hearing is a good look for the industry. And governments love good looks.

Well I've got news for the governments. The mess that the residential building industry is now in, will cost the economy far more now than it would have if disputes had been made fairer for home owners by having a proper definition of defect.

******

Building Defect

Quite often, one or both sides will be offered (and opt for) a building authority appointee involved in the dispute... all part of the system we have.

I have been involved in four cases where building authority appointee building consultants were involved (one with consent of both sides and with no right of redress); and all four of these appointee reports were (I say) all unfairly and heavily weighted in favour of the builders. In one instance the appointee price was 52%, the second, just 5%, and for the third and fourth cases, less than 25%... of the amounts which a duo of registered building practitioners, (a builder and an architect), both knew was necessary to bring the homes to a workmanlike standard... to a standard expected in the industry and stated in the building act and contract as a requirement... not band-aided as was actually proposed by those appointees. They were in effect opposing the home owners, because their definition of defect was pathetic.

It was so obvious to me that these consultants did not have a proper definition of defect; and yet they were still prepared to deny actual defects alleged by me on the grounds that there was no failure to date. They would not consider time and life-expectancy.

I think that these 'appointee' consultant reports were supposed to make the owners feel that their expert was over the top; and that this added pressure would make them bend a lot more than the builder towards the quite ludicrous amounts assessed by those building authority appointees. The reason for the continual re-appointing of these same consultants by the building authorities, as far as I can make out, is that they get results; they help achieve the aim of the building authority and/or government to resolve the dispute as economically and as fast as possible, for the good of enterprise and the economy.

But this is certainly not for the good of the houses or the home owners… and as I will later show, not for the good of the long-term economy either.

And the lack of a definition of defect is what allowed (and still allows) this to happen… over and over again.

So I say… Let's stop the nonsense and make it fair once and for all.

Two of these building authority-appointed consultants were negligent in their duty to the tribunal in more than one way and in more than one instance regarding major items, but unfortunately in those instances I was thwarted and not given the opportunity to expose those blatant misdemeanours, by not even getting the opportunity to be heard by the member in charge of the court hearings: all because the home owners wanted to hurry up the result. (And there was a gag each time). A third building authority-appointee was found by the court in a hearing to be grossly inept in his pricing by the presiding member. And that particular consultant was by far the fairest of those appointees: but he failed to scope his estimates to achieve a reasonable life expectancy for each component; and as a result was way under the appropriate price.

I believe my building science knowledge was largely responsible for the temporary removal years ago, of more than one arbitrator... for bias and ineptitude. The opposition solicitors agreed too; and got their clients (the builders) to give a serious offer there and then; and so the cases were terminated, arbitrator and all. I believe at least two of those arbitrators eventually worked in some capacity for the sole insurer and one eventually for the ex-building commission. It seems a likely transition and the way that bureaucracies tend to evolve. People already in power often seek similar positions of power when a system is changed... so that old habits tend to prevail for long periods of time, just as insurers now seem to follow along the path that was laid down by the first insurer, the government-owned Housing Guarantee Fund Limited.

And way back then there was also no definition of defect.

The system at present lets Building Companies fight home owners with bluffs and negating building consultants, whilst VCAT personnel presiding at Mediations and Compulsory Conferences pressure both sides without a definition of defect. This must favour builders because many actual defects can be swept aside as mere maintenance items. Then the legal system (with its gags) comes into play, and the general fear of the legal system does the rest. The result is that little or no inquiry is made into the builder's quality of workmanship, even where a Builder repeatedly appears in VCAT against a succession of disgusted clients. These builders are protected by the privacy afforded in the mediation process. They are also canny enough not to expose themselves to a hearing, and because of this, the public does not find out how bad their workmanship really is.

Welcome to Common Law.

When things go belly-up, I think the vast majority of businesses would do just what the struggling builder does when the claims against them exceed their (limited) liability. They would go bankrupt.

The vast majority of them are not going to dig into their hard earned (private) assets just because of one bungled house (or one bungled deal). They might be proud... but only to a point... that's what limited liability implies as I see it. That's business generally! Since 1996, we've seen fairly large building companies and one building warranty insurer go bankrupt... even in these boom times. Builders and insurers are proud... but only to a point. And the more important the dollar becomes, the less pride there is in the company name.

Bankruptcy is not a criminal offence as it once was in England at least, but smaller businesses and other creditors must still wear the hurt as they did in those olden days.

Welcome to Company Law.

Now I say it would be ethical to pay back in full what you owed your creditors. I can see how it can happen when people one layer up don't pay you – it's like the domino effect – but very few stop the domino effect and pay back what they owe over subsequent years. I personally know only two. (These two stopped the buck-passing to those next in line and stood their ground... and it took them years. But eventually both got going full bore again fortunately... and the important thing was that nobody else got hurt. They were brave and very ethical people, but unfortunately a rarity in business in today's society.

Businesses generally regard this concept, not as being ethical, but as being 'suicidal'. So how can those businesses really claim to be totally ethical if they think like that? It's a sham.

And businesses don't necessarily charge what would be a fair price for a fair days work, when the going rate (the opportunity cost) of similar businesses is double that fair rate. They simply let supply and demand of the marketplace rule their decision. They basically get as much as they can. This is what happens during boom times in our community.

Welcome to Business, Business Ethics, Private Enterprise & Consumerism.

And how many poorly built homes does it take for the governments and building authorities to admit they have made a huge blunder by relaxing all of the rules without sufficient checks and balances. It could well involve over a million homes Australia-wide already.

Welcome to expanded privatization and the freeing up of regulatory controls.

AND... do not lose sight of the fact that large firm builders (in fact nearly all large businesses) are large for basically one reason, those in charge want to become richer. But it's not just business

Just take a step back to see the size of most of the houses built for young couples today, and you realize that people are basically out to get what they can when the opportunity arises... and low interest rates means increased opportunity. In the early 80's, the average home was about 13 squares (120sq.m) with 22 squares (204 sq. m) thought to be quite spacious. Now 22 squares (204 sq. m) is average and homes are being built in surprising volume right up to 40 squares (371sq. m) and more.

Welcome to greed.

Society lauds and rewards fast and successful business growth with special awards. The more people employed by a new business, the better it is overall for the economy and the community it is said. Sometimes that may indeed be the case... but often it is not the case at all. The huge success of one business is often at the expense of several other less competitive small businesses nearby. There is often a drop in the overall employment. One big business instead of (say) five small businesses means about 4 less Chiefs and perhaps only one or two more Indians. And less chiefs means less expertise.

So the growth of firms can sometimes be detrimental to the community at least in terms of expertise and experience. It can dumb down the community by reducing the number of business- savvy people in the community. In the short-term at least, it will lower expertise.

So greed can have an effect on expertise and experience, and because of this, the quality can be substantially reduced, as has occurred in the residential building industry recently.

Building Defect

Even with the stabilizing factors of increased availability of credit, less economic volatility, preparedness of the banks to lend more, sustained fairly low interest rates and inflation; it worries me that home owners today seem to have the greatest risk ever of holding onto their main asset…

Because defects built into their homes are also running at a record high.

******

The following graph tells how important the home asset must be to home owners, where owners are now borrowing something like 1.5 times their total income compared with just 0.3 times their total income in 1970.

And they are borrowing against the equity they have in those homes, their main asset.

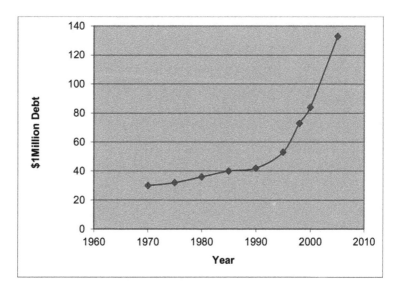

Source: OECD, ABS Thomson Financial, AMP Capital Investors

Graph of Household Debt Levels as % of Household Disposable Income

By becoming larger, builders benefit from economies of scale and gain more bargaining power. Some of these savings can be passed on to their customers to improve their edge over the opposition. But there are also associated costs to many more home owners.

The one-to-one contact with the experienced builder is vanishing, in many cases being lost altogether; and alongside that, the experienced eye is no longer on site looking after your house.

You and your house; to put it bluntly, are simply not important enough for the experienced boss to actually supervise the building of your house, and so things can and do go wrong.

Sure, the company may give you more attention as regards colour schedules, equipment inclusions and the like, but not the expert quality supervision. The top dog may sign your contract and take your deposit, but from then on, you're lucky if he has any more to do with your project. You may not even meet him. He may never even visit your site.

Which is why, when a dispute arises with a smaller (unlimited) builder, he usually shows up in person, whilst when a dispute arises with the larger company builder, the boss is often represented by a manager or equivalent of the company. It's the same principle as above. And when you point out to the supervisor defects you discovered at the final inspection, the problem is handed over to the maintenance department, not the boss. So you get maintenance carried out, not rectifications.

******

On two occasions in my experiences with big building companies, fair-minded maintenance personnel were replaced by not-so-fair personnel, with the result being a reduced list of previously agreed items that were to be rectified. In one case, rather than pay what VCAT had handed down, the builder took a full hearing decision to a higher court to argue, (I believe), about the word 'substantial'... and lost. This is now another precedent.

It seems that there is an ingrained reticence by builders to pay the full amount to rectify the mistakes that they make.

When the chips are down, they invariably resist the temptation to be fair.

So if you are in dispute, you may well find yourselves taking on builders who have the following business maxims:

- *If you hang in there long enough the clients will usually give up,* and
- *If you act as if you're being reasonable, the clients will usually give ground,* and
- *If you begrudgingly give enough to avoid a VCAT hearing, why give any more,* and
- *By using a gag you can probably stop the community from finding how bad you are as a builder.*

Because at the end of the day, business is business! And as the insurance solicitors say...

Welcome to commercial reality.

In other words it pays to fight home owners. I kid you not... that is what they say.

If builders and negating consultants are repeatedly permitted to deny many of the legitimate defects that are found; what does that say about the system that allows these perpetrators to offend and then re-offend? It is repulsive to me, that many people masquerading as building consultant-experts can be repeatedly unethical and/or grossly inadequate at their job... and still be paid a lot for it... by the clients who have most to lose.

Building Defect

So home owners are very unlikely to be offered what I consider is a reasonable amount of money... sufficient to return their house to a workmanlike condition... until a definition of defect is established.

But I still say persevere if you wish to do your best. Just get a capable team together and be aware that there are quite a few precedents that may rear their ugly heads to rob you of what should in my opinion be your basic rights. Such is the case when using the word warranty.

Already there are several firms who attend to whatever the builders agree to fix and almost certainly the way the builders say to fix those agreed items, and larger firms have a fleet of maintenance teams that do likewise... **so that VCAT will not be involved**, simply because home owners (not knowing of any better method), agree to this approach which seems fair at the time... AND... **it's the easier path to take**.

In the name of MAINTENANCE many of those repairs will be pure band-aids such as this typical use of silicone

Silicone was applied more than once to fill the gaps and is failing or has failed already after just 4 years.

Other fillers are also used in different situations inside and out, with similar failure periods of short duration.

But when these houses reach 6.5 years of age, all of a sudden the builder's response is…

"Sorry, the warranty period has ended".

And all those owners have from then on are maintenance - prone homes. End of story.

What so few owners realise is that they have been putting up with mere maintenance, when rectifications were warranted (sufficient to cure defects): work that would outlast the warranty by a considerable time margin and

Building Defect

reach a reasonable life expectancy and not just last a few years if lucky.

Chapter 9

The Ex-Building Commission Failed to Stop the Declining Standard of Building Work

This chapter aims to show all of you the degree of thought necessary to cure a particular problem in the industry... this one a safety issue that could affect any one of us at some time in our lives. And a proper definition of defect should kick-start the healing process.

******

The ex-Building Commission acted via its guideline '*What you need to know about maintenance of balconies*' and a summary of this guideline '*Let's keep balconies safe*' appeared in its regular publication '*Inform*'. This was possibly in answer to the failures of a few balconies such as the one at a rowing club in Corio near Geelong in Victoria. What seemed to be the gist of the cure proposed by the writer of the article, was vigilance by owners.

Now I agree that vigilance is a very, very important factor - but there was no mention at all of the builders of the balconies or the building surveyors and building inspectors who passed these balconies at plan stage and whilst being built. And there was no mention of those who played a part in specifying the construction of the balconies (people such as architect / structural engineer / draftsperson / builder).

Liability associated with using and/or condoning the use of non-durable timbers in an exposed situation... timbers that often do not reach a reasonable life expectancy was not mentioned at all in the article.

Instead there was a warning in the guideline of being wary of those structures constructed without a permit, plus the mention that inappropriate timber species were used in the construction of quite a few decks and balconies... *"during a period when balconies were becoming increasingly popular but safety threats were not understood"*.

Safety threats *"not understood..."* baloney!

Ignored was the word that should have been used. And negligent would have been the correct follow up... but not a mention... so as not to lessen people's faith in specifiers and builders perhaps. Well that's not good enough... and nowhere near tough enough!

Since the 70's, nearly everyone in the building industry has known that hardwood and oregon used outside for pergolas and fascias and deck framing timbers were going to rot and be likely to become so rotted as to become unsafe usually within 15 years, sometimes in as little as 7 years. I've seem substantial rot in decking joists in just 4 years and the deck was built by a builder in about 2002.

The problem was that very few builders used CCA treated pine, for the simple reason that it was considerably more expensive... and the owners may have been frightened off by the higher price.

In the article, there was also no spelling out just what "inappropriate timber species" were used... and no mention that these same materials are still being used today (although less frequently)... again perhaps so as not to cause sudden harm to businesses that sell these non-durable timbers to builders and owners, and as usual to protect enterprise.

So why was there no mention of prevention of collapse at the beginning of the article?

Balconies have been commonplace since the late '70's. Unseasoned hardwood and oregon were the most commonly used timbers. But the summary of the guideline does not specifically mention hardwood at all. And by the way, even with maintenance, hardwood and oregon often start to fail well before 10 years is up (often slower if the balconies are situated on the north or east side of houses). Even then I've seen completely rotted framing in east decks just 6 years old.

To say that *most* well maintained balconies last 20 years is fine to a point, but what about the ones that don't last anything like that period of time... surely they are far more important. There are a multitude of balconies that are unsafe in well under 10 years.

And failure or potential near-failure at under 10 years of age is a negligence-caused-defect... caused (in part) by

Building Defect

most of the building practitioners involved in their construction.

But the ex- Building Commission did not spell this out in detail. Twenty years almost sounds like a reasonable life expectancy. Maybe that's why they picked that figure. But it is what was not said about the negligence aspect that worries me.

Remember what I said about hidden agendas... well the absence of such a comment supports a theory of mine: that by leaving business out of the argument, there will be less claims made by home owners. Make owners think that they are partly to blame and the thought process of blaming the builders and specifiers may never arise.

So I say we need to fight this protective apathy. Maintenance has a part to play but so too does negligence. We must get at the cause, not treat the symptoms. **Many of these balconies are defective and must not be treated as merely maintenance problems.**

Even today there are decks, balustrades and quite a few balconies being built using oregon and unseasoned hardwood... many covered by inadequate top edge plastic strips or floor tiling on inferior substrates... sometimes even particleboard. I have seen a larger builder admit to this potentially unsafe type of construction in a Building Industry magazine, in order to promote a better product.

Today, lots of deck structural members (most not totally durable) are not provided with top edge protection... and what's more it seems that most of the top-edge protection when used is also defective because it does not have a bitumen content to seal around the nails... because it's cheaper and looks good. Inadequate top-edge protection was not specifically mentioned in the ex-Building Commission article... and just as for specifications that are no more than a list of materials, it served to dumb down the industry that little bit more... and home owners (and their decks) will be the losers... again.

And I understand that LOSP treated pine (as distinct from CCA treated pine) can develop substantial core rot too if not adequately end-treated and top-edge protected. But does anyone in authority care about that?

The *Inform* article made little or no mention of how the builder / architect / certifying building inspector / relevant building surveyor could insist on measures to prevent the likelihood of a premature failure, or perhaps stop it from ever happening. The emphasis was solely on how owners could help to avert a possible disaster. The emphasis was similar in the ex- Building Commission's Guideline summary.

When is someone in authority going to take full responsibility, and prevent such structures being built from non-durable timbers from now on? For starters it would be a green and a long-term economic initiative, and would also stop the wastage of labour and materials when poorly-built balconies needed to be re-built.

And when are non-durable timbers going to be adequately protected in exposed situations... so that they can reach a reasonable life expectancy? Chapter 17 is devoted to this question.

And when are hardwood fence rails going to be phased out? When are red gum and cypress pine posts going to be required to be protected in the ground for sustainability reasons?

When are we going to really tackle the problem of sustainability in the building industry?

The next example shows what I mean about oregon still being used...

Building Defect

This is one of at least 10 units with balconies overlooking a marine mini-harbour. There are almost certainly hundreds in Australia similar to this. Note the orange colour of the floor joists through the latticework... that's oregon... and oregon is not durable. So I say that these floor joists are defective because they are likely not to reach a reasonable life-expectancy... at least 25 years is fair I say... and so does N.A.F.I. (a top or perhaps) THE top Australian Timber Authority.

The following pertinent questions therefore come to mind:

- How can the owners monitor the rot in these pictured floor joists (they are over water) and why should they have to?
- Do many owners even know they are advised to monitor their balconies?... and

- Why should the builder not have to replace these non-durable joists... with durable ones... now?... and
- Is the building inspector also to blame for permitting this to be built?... and
- Is the building surveyor also to blame for allowing oregon to be specified?... and
- Are the specifier(s) (architect, draftsperson, structural engineer, builder) to blame?

You may be interested to know that you are still permitted to use red gum and NSW cypress pine stumps (at least when re-blocking older houses). And yet structural elements are mostly supposed to last at least 50 years. NSW cypress pine stumps can possibly make it to 50 years with sufficient drought periods, although 35 years can also be its lower limit... and I've seen this timber rot in as little as 3 years due to a yellow fungal attack peculiar to this species, the fungus being inherent in some of the product. Much of the available red gum would be lucky to last 30 years without in-ground treatment. So who owns this problem?

Under my definition of defect, the use of such materials would be regarded as negligent, and a defect on the grounds of not being fit for purpose... to last a reasonable lifetime.

Perhaps the lack of mention by the ex- Building Commission of possible liability of building practitioners avoids alerting the public to the idea that the rules, (loose enough to allow some decks to be built from inferior non-durable materials), should be re-examined.

And that might spell Trouble (with a capital T).

The failure of the Corio, Victoria balcony was largely handballed over to the owners in a similar way that the CSIRO document 10-91 'Maintenance of Footings' (and its inferior successors referred to in soil reports), often has the effect of making owners think it is their fault their building has cracked, when they should often be pursuing the builders of their homes to rectify wall cracks caused by the builder's (and/or the privatised relevant building surveyor's and/or the structural engineer's) disregard of recommendations in the soil reports. Several times I have seen this happen. Once a requirement of minimum 2200mm depth of footing by the building surveyor based on a soil report recommendations was passed as adequate by the building inspector at just 700mm... because an unstamped plan copy issued prior to the altered upgrade by the building surveyor was permitted to be used on site... a bigger builder who did not follow the rules established decades earlier... that a stamped plan must be on site at all times.

I wondered if the failure of the deck in Corio could possibly have been due to the fact that the live loading code requirement for decks was up-graded (or wrongly graded or down-graded) following the realization that people's habits have changed and more people are likely to squeeze onto balconies and decks nowadays for a few drinks or whatever. Why was it not discussed in the article as a very real possibility for the failure of decks? Again, my theory is that it may have led people to realize that there are people other than owners who are liable for ensuring that the deck is required to be adequate for the intended use (designed for appropriate live loads).

I wonder if the collapse could have been avoided by a change of use regulation as well as by vigilance. Just what was to blame was not clear in the paper I read. In fact I would like to know if reasonable vigilance could have prevented this Corio accident at all. Rot is difficult to see sometimes; particularly when it starts from the top or the paintwork is dark. Maybe it was inspected, and the building consultant did not inspect it thoroughly? And what exactly constitutes reasonable vigilance?

And even if the Corio balcony collapse could have been prevented by vigilance, vigilance alone is definitely not the answer for all balconies, as I have pointed out. Like one's health, prevention right from the outset has a major role to play… and I say that for decks and balconies being built right now (many with hidden non-durable framing), prevention from the outset has a far greater role to play than vigilance.

It seems that very often, we the public are not told even the basics. You get told "what you need to know and no more". As spelt out earlier, specifications no longer describe what workmanlike means. So those of you who are reading this will now know what to question.

And that may spell TROUBLE (all capitals).

Now for existing older 'at risk' balconies – yes – vigilance is of the utmost importance. But so too may be the inadequacy of the design. So too may be the reliance on building consultant advice that nothing is wrong. Some consultants inspect whole houses in just over 15 minutes... Yes that's all the time some of them are on site... amazing when you see it (and I've seen it twice at pre-auction inspections). I saw one miss the only major defect in a 1920's 30 square Canterbury (Melbourne, Victoria) house. He took just 17 minutes to complete his inspection.

So what about the balconies being built right now and those still under warranty which (I say) have to be built to last a reasonable amount of time... at least 25 years? Many of them are still being built using materials that are not fully durable and which can be affected by at least excess moisture and become unsafe due to incorrect installation and also cannot be adequately inspected.

I have seen several exposed timber and tiled timber-framed balconies and decks built since the article, that are not going to reach a reasonable life expectancy. So the article and the advice has definitely not filtered through to every builders and specifier.

I also think that the reliance on whether or not a building permit was issued for the balcony or deck as being sufficient safeguard against collapse shows a total lack of realisation as to just how many decks, balconies / balustrades (and generally most houses) out there, are inadequately built.

Building Defect

I really think that the authorities simply do not know how dire the situation has become out there... on so many other issues apart from decks and balconies.

And each house with a balcony that has been given a permit has had a designer, a specifier, a builder, a building surveyor and a building inspector. Some had a structural engineer, others also an architect. And even with all of these responsible building practitioners involved in the process, the average house completed since 1997 has far too many defects still waiting to be discovered, including many potentially unsafe balconies and balustrades.

Just as worrying to me is the realization that the average building consultant inspection report uncovers less than 30% of such defects... and in more than a few cases... nil defects!

Two building consultants in my sample of 22 follow-up reports missed rot in balustrades, one missed rot in the actual deck substrate (particleboard)... all three were less than 5 years old. One deck was completely rotted in at least 20 locations. Another built in 2001 and elevated too, had rot in most floor joists, the main beam and ends of most balustrade rails... they were all made of oregon or hardwood. Another 2002 example had rot in all the main beams and several completely rotted floor joists. It was irresponsible of the ex- Building Commission article to single out just the 80's and 90's.

Many under-warranty balconies and decks are potentially unsafe before the warranty runs out. So please let's not say that they are OK because they have a permit and certificate of occupancy (final certificate)... because a lot are far from OK. Some are a disaster waiting to happen. The successor to the ex-Building Commission needs to get out there and see what is actually being built... and think long and hard about the future!

Builders (in fact building practitioners generally) are often liable and should be made to pay for their mistakes... then and perhaps only then will we start to have some real care taken in the building construction process.

A few well thought-out regulations might also help.

So let's stop this pussyfooting around the periphery... and in so doing, protecting the irresponsible offenders responsible for this fiasco. Let's not wait for a case that makes the courts decide. Let's not wait for a death or two... let's regulate to avoid these possibilities.

SO... I request an immediate up-grade regarding balconies... to include the liability issues...for the sake of everybody.

So when a Guideline starts with *"When the construction of a building is complete, the building owner is responsible for its upkeep and maintenance, particularly its safety features"*... all I see is authorities with blinkers on... with no idea as to the actual building problems out there, no idea that approved and completed inadequacy is commonplace, and no idea as to the very large number of building problems actually out there.

I recently talked to a man who fell through his 4-year-old tiled elevated deck onto the deck below... and there were similar adjacent decks waiting for the same accident to happen. He didn't die but damaged his leg and ribs. And it was a real possibility that he could have died or become bedridden for life... so what about the other decks of the adjacent units? I believe the substrate was too thin and therefore too weak for the span between the floor joists.

When is someone going to do something about all of these problems?

Do you see why someone needs to speak up?

******

I think that's enough on balconies, but it needed to be said and needed to be an in-depth appraisal to show all of you that the authorities have unfounded faith in the quality of work being built out there. And I have no doubt that the article was written as a genuine attempt to care for us all. I just wasn't anywhere near good enough and was too protective of business.

******

Next I turn my sights to building warranty insurance, because that also needs an immediate up-grade so that it covers home owners (and subsequent owners) to the level of their expectations... even if the policies have to cost more... which I say is debatable. (They seem to have had it very easy for quite a long time now).

I am about to reveal some more hidden secrets, that once again, nobody seems to have challenged.

Incidentally since the first edition of this book, not only (in Victoria, Australia) did the name of the top building authority change yet again to The Victorian Building Authority, but the building warranty insurance was again taken over by the government insurance authority Victorian Managed Insurance Authority... and is at least as mean if not meaner than any insurer of the past.

There have also been a 'tranche' of building reforms in Victoria, the sum of which has resulted in an overall more repressive industry particularly as regards home owner rights in disputes, the DBDRV and VMIA insurance forums ensuring that builders are protected as much as possible, no matter what their misdemeanours.

This is not just my opinion... but virtually all consumer groups have been largely ignored as long as possible in Australia... with the latest group of suffering individuals being apartment tower owners... deserving of another book.

Chapter 10

Domestic Building Warranty Insurance Covers Far Less than You Need It to Cover.

Building warranty insurance cover is not what it used to be. It is still of great benefit to those home owners who are unfortunate enough to have had their builder go bankrupt or walk away during the construction of their houses. But after just 2 years there is no cover for anything apart from structural defects, due to changes made in July 2002. In my opinion, precedents also need to be removed by legislation to create a fairer playing field, even if their removal means greater premiums. Insurance cover for home owners simply does not meet with their expectations.

In July 2002, following the crash of HIH (part of the FAI Insurance group), the government got very concerned that they would have no more insurers interested in insuring houses at all, and the government could be seen as a failure. The crash sent shivers down politicians' spines. And when this sort of thing happens, you get action… and the action in response to this sort of collapse looks after business first and the individual a distant second.

And that's just what the government did - they got together with the remaining insurers and came up with a proposal that doubled the amount of cover (from $100 000 to $200 000). It seemed to everybody a very generous move. The structure of each house would still be covered for six (6) years, a little down on the previous six and a half (6.5) years. That wasn't much of a reduction. So it still seemed as though everyone was a winner.

Well I've got news for you... we've still got building warranty insurance... but it actually covers far less than it used to as I will soon show.

The revised insurance cover was a real boost for owners of houses still under construction with bankrupt or walk-away builders... perhaps also for owners of very large houses.

But for the vast majority of home owners, these new policies were not generous at all... because, as a trade-off, whilst the cover for structural defects was set at 6 years (reduced only slightly from the previous 6.5 years), the cover on all other defects was reduced to just 2 years, replacing the 6.5 years cover on everything of earlier policies).

But what we were not told was that vast majority (about 90%) of defects are non-structural.

And so you, the public, have been denied, in effect, over two thirds (69.23%) of your non-structural defect rights, by my calculations. This gives you a cover of $2 (1 - 0.1 \times 6/6.5 - 0.9 \times 0.69) \times 100\%$ which equates to 57.34 % total of what you previously had. This equates to a

REDUCTION in insurance cover of 42.66%. Those of you who are mathematicians might have noticed that there was a factor of 2 included. This is to allow for the doubling of the maximum total liability of the insurer on each policy.

So insurers got a fabulous deal... and far more than just the 42.66% drop in cover. They were saved the enormous hassle of chasing the builders. **Home owners must now pursue their builders (to bankruptcy if need be), before even approaching their insurers.**

This must have been an enormous saving to the insurers. How the Government could permit this is beyond me, unless the remaining insurers knew that they had the government where they wanted them, knowing that the government did not wish to return to a government-run insurance scheme. And so the remaining insurers had a field day and lowered their risk.

With these obvious savings to the insurers, it seems to me that the additional cover I propose should not make warranty insurance much more expensive than it used to be prior to the changes. And by the way, $200 000 cover is not always sufficient in my experience... for the considerable number of home owners who own a *'lemon'*.

So to all home owners: act promptly (within 2 years), or you may not have the insurer to fall back on for most of the defects in your house, should the builder go bankrupt. I've included another 2 graphs so that you can see why you had better act promptly.

A picture leaves a more indelible impression than words... and so I have included some graphs to explain what has been a considerable reduction in home owner rights as regards domestic building warranty insurance.

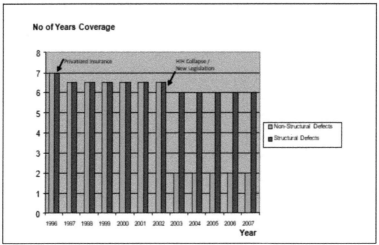

Graph – Insurance Cover for Defects

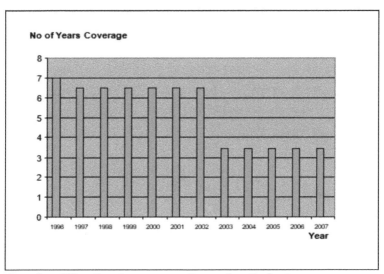

Graph – Effective Insurance Cover for Defects

And the situation has not altered since 2007 by the way.

Many complications can occur as regards the rights of home owners and buyers in a real estate venture concerning a house (or house with additions) under warranty. Seek legal advice on your rights, because this book is not about specifics... and authorities often change the rules. So read the insurance policies and the building contracts and find out.

When inspecting houses still under warranty, building consultants should also be aware that if the buyers are depending on the advice in those reports prior to placing an offer; then considerable liability may be attributable to those building consultants, particularly if the buyers are not alerted to all of the defects in such houses... and do not realise until too late that there are far more defects than they were told about.

Once you have bought your less than 6.5 year-old home, then you have only until the end of the warranty period to make a claim for any defects (that I suggest only a capable building consultant should be asked to inspect)... but only against the builder don't forget. The insurer only covers most items for just 2 years and structural items for 6 years. And history tells us that quite a few builders do go bankrupt for a variety of reasons. It's tough out there! **Please read your policy to find the deadlines. They may be different already.**

Historically, insurance in the building industry was government-run and started with cover at a meagre $40 000. That was often nowhere near enough. The government and HGF (the only insurer), eventually realized that this was nowhere near enough; and increased the amount of cover to $100 000, which for most houses is still adequate today... but not for all... and certainly not for calamities which occur during the construction period.

I have however been involved with at least 12 houses where the amount to rectify what was defective (to the contracted workmanlike quality) was well in excess of $100 000.

Now the limit is $200 000 (from July 2002 onwards), and despite the fact that this amount is large, it is my opinion that even this is already not sufficient for some of the very poorly built houses, or for larger houses being built. (Reasonable legal and other legitimate costs were awarded on top of the amount of cover by the way... but there is now one very important precedent (Isaacs versus VERO) which re-defined what costs are considered reasonable).

Owner-Builder insurance is quite different in that it does not exist unless the house is sold prior to 'X' years elapsing since the completion of the house. It was once 7 years but may well be different now to tie in with current insurance limits perhaps.

A report (called a 137B report) is also required just prior to such a sale, so that buyers are aware of what defects are in the house (or what defects were band-aided) prior to agreeing on a price. The insurance policy then covers defects NOT included in the 137B report for the remaining

Building Defect

time and with typical terms and conditions usually written very small indeed.

You can imagine the complications that can arise if the building consultant who carried out the 137B report found only a small percentage of the defects. Unfortunately this is a very commonplace scenario in my experience, (and includes some extremely deficient reports by some building consultant firms that I understand are recommended by some authority in the building industry). Often when you realise that your reporter was inept or that he was reporting with a set of disclaimers that meant he inspected only half the house, it is too late to claim on your insurance.

To this point I have been concentrating on new work; basically new houses. But the same applies to several other types of works besides new houses – namely extensions, additions, alterations, renovations and re-located houses. Some smaller types of works do not require insurance or permits, but the workmanship in some of these jobs can be just as poor as that found in new houses, sometimes worse. (One of my clients had 4 such minor jobs carried out by different tradesmen… and all 4 jobs basically had to be re-done).

To have your houses rectified, owners of new homes and new additions must fight the builder first and the insurer will only become involved and take up the slack when your efforts against the builder have come to nought… and take it from me as a near certainty; both will fight you in an attempt to deny you your rights.

Chapter 11

Builders and Insurers Fight You Because It Pays to. So Many Home Owners Just Give Up

The Housing Guarantee Fund would not have used negating tactics against home owners if they continually lost cases. They fought home owners because they usually managed to reduce claims against them sufficiently to come out ahead, even with all of the legal and expert costs involved. Builders and Insurers (VMIA in Victoria) today do the same thing for the same reason.

They do this because:

1. It is so easy for them to warp the definition of defect to mean maintenance, because there is no official definition.
2. They are permitted to say that they will rectify defects when they are merely going to replace the failed short-cuts already installed with a mere repeat of the short-cuts... (It's called a 'band-aiding').
3. During the mediation process, most people take the easy road of allowing the builder back to rectify (only some) of the alleged defects that the builder

couldn't build properly when being paid. (And those items agreed to are usually just those that can be rectified cheaply). People also fear the legal system.
4. Most home owners employ building consultants who are not expert or capable.

And so something like 70% of home owners give up... (That's 70% of the mere 4% or so of home owners who enter into a dispute in VCAT in the first place). So I suppose the government thinks everything is going smoothly... and that nearly everything else must be well built. What a poor system... because it seems to me that most homes and additions built since 1996 are worthy of a dispute. I have seen very well built homes, but not many.

Another group of home owners, perhaps the vast majority of them, allow the bigger builders to repeatedly maintain their mistakes, (under the guise of caring, via their attention to paint blemishes often found by the builders' quality control people and supervisors). So why didn't those quality control & supervisor people find the other defects in those houses?

Perhaps many of the owners of these homes think that there are no other defects in their homes other than paint blemishes and the like; because they are too trusting, or they are perhaps too hocked to the hilt to find out before the warranty period is over. How could people know, if they do not have any description in their specifications on what labour is required to achieve a workmanlike job. And authorities are certainly not sending out alerts.

But quite a few more defects will show up well before they have reached a reasonable life expectancy. Often these defects will become obvious after the warranty ends.

For many home owners it will be too late to do something about it, and our community will be left with a rather large and growing repair bill. But some the defects will show up before the warranty period ends. But as I said earlier, these owners will still have to fight their builders if they want the defects rectified and not just band-aided via the term 'maintenance'.

Here are examples of negating techniques used by builders and negating building consultants; designed to put off home owners and reduce their resolve and faith in their own building consultants.

TYPICAL BUILDER RESPONSE LETTER (with my comments in brackets)...

- The alleged defect is due to normal wear and tear (maintenance item they say)
- The alleged defect seems to be performing O.K. (calling your bluff to prove it)
- The alleged defect could not be found (delaying tactic hoping it will go away)
- The alleged defect was apparent prior to the final payment and is therefore not claimable (the Insurance company building consultants use this one without knowing whether you saw the defect or not)
- The alleged defect has been approved by the Certifying Building Inspector (all have been, but

this person is actually not being paid to look at the quality of the work)
- The alleged defect will be investigated further and the manufacturer requested to comment (the builder should know how it was required to be done and may be just using a delaying tactic or possibly does not wish to give his opinion)
- Your building consultant seems to have exaggerated the situation (this one related to missing roof fixings... and there were actually 20% more than I estimated)
- I am not aware of any regulation relating to the alleged defect (quite often there is no specific regulation that directly controls avoidance of particular defects other than the requirement that the work complies with the National Construction Code (formerly the Building Code of Australia) and the Domestic Building Contracts Act which together require workmanlike building construction. It could well be a lack of building science knowledge).
- Technical aspects relating to this alleged defect are being checked (not saying no to make you think that they may agree but usually just another delaying technique).
- The item seems passable to us. (In one of these instances it was the mortar strength, and the owners had the cement content tested and it was proved to be inadequate and under requirement by over 50%).
- Proximate cause unknown (means anyone could have caused the defect)

Again, there are many more I could have listed.

I said earlier that your specification was likely to have no general or workmanship clauses contained in it? That keeps you, the owner less informed... and makes it just that little bit easier for negating building consultants to call your bluff about what is considered workmanlike. Bear that in mind when you receive their responses... because it helps a lot to keep your resolve if you can see through their tricks.

Now whilst it is not a requirement that the builder be permitted to return to carry out the necessary rectifications, insurance policies and some authorities seem to take it upon themselves to try to make this happen, probably because it saves the original builder overheads and profit margin costs that another builder would charge. But if the original builder can be shown to be suggesting band-aid repairs of defects (by agreeing to the cheaper items and denying the more costly ones), then I strongly suggest you do not go down this easier path, or you will simply be saying yes to a repeat of what has already failed because the original builder did not supervise the job adequately. His consultant tries to fool you by not saying (in detail) how the builder should fix the agreed defects.

This is what allows the builder to band-aid your repairs as discussed in Chapter7.

Most of the items home owners claim are agreed to because there are cheap ways the builder can use to get you off his back. But there will be no detailed scope of works. There is such a strong trend towards rectifying only those items that can be addressed cheaply, that I have come up with the following theory...

If the builder or building consultant for the builder or insurer cannot think up an easy method to rectify an

alleged costly defect, then every reason under the sun is brought forward (from the bag of dirty tricks) as to why the alleged defect is not a defect at all.

Sometimes I think they have been trained to respond this way. They all seem to act in a similar fashion. But even if you don't accept the builder returning to the site to rectify the defects agreed to, the opposition may still give you an offer (a lesser one than you hoped for) in order to end the dispute... if you are lucky. I know of three offers that were over 60% of what was needed to rectify properly, but most offers have been about 30% or even less.

So if you are thinking of taking the easier route by accepting an offer made by the builder or the insurer, then you should consider this... because it is going to happen... you will be asked to accept a gag.

GAGS

My advice to everyone is: try not to accept any form of gag. You simply place yourself at risk of a potential libel or slander lawsuit. If there was no risk then why are gags used in the terms of settlement? Just by telling your story in too much detail, you could place yourself at risk. To me this is just another form of bullying... by the system for the sake of enterprise at the expense of home owners. It happens in almost every out-of-court settlement.

Often during the mediation process, home owners are eventually offered something half-reasonable. (I say half, because it usually is about half... or less). But

invariably such a compromise settlement tendered comes with strings attached such as a gag.

The easy road is to accept the gag and be bullied one last time... and most owners do this.

With our present system, the actual amount you end up with, (after paying excess costs), by going to a full hearing; may not be very much more than you were previously offered. But it will NOT BE CHEAPER for the builder or for the insurer if this occurred, because it often happens that they end up paying most of the costs IF you have a capable team.

So there is more left in your bargaining kit than you might realise. Unfortunately the builders and insurers have until the day of the hearing to give you that reasonable offer... and most home owners don't bother to get there because more and more costs will build up as the barristers prepare your case for the hearing. Disputes are tough.

If a gag stops a builder's business from being threatened, then that may well be good for the economy! However that builder can repeat the shoddy building methodology that caused your poor quality home in the first place – over and over again. (And that's worse for the economy). Just as long as the dollars coming into the business exceed the dollars going out, all is well... and pride can continue to take a back seat. To me, that basically sums up much of the business end of the Domestic Building Industry today, particularly when there are fewer and fewer players, as bigger building firms continue to grow.

A building company COULD change its systems to ensure that the houses it builds no longer have so many defects incorporated, but then it may not be quite so competitive, because the builder would need to police the tradesman more and pay them for the extra work they used not to do. Maybe it would be better just to ride out the lawsuit bumps and let the system, the maintenance teams, the inept building consulting and people's fear of the legal system continue to placate or put off the vast majority of potential legitimate cases.

All in the name of a competitive privatised industry… '*All in the name of progress*' as they say.

If the governments or authorities actually get real and decide to up-date contracts, perhaps they could also update Section 32 involved in the sale of real estate.

As well as the requirement on the vendors divulging what work has had permits in the last 7 years, what about adding the requirement that the vendors divulge **WHAT WORK HAS BEEN CARRIED OUT WITHOUT PERMITS IN THE LAST 7 YEARS (OR THE PERIOD OF TIME THE CURRENT VENDORS HAVE BEEN OWNERS OF THE LAND) WHICHEVER IS THE LESSER.**

Terms such as *'performance',' workmanlike',' specification'* and *'defect'* have none of them been adequately thought out… and home owners, home owners-to-be (and the economy) will be the losers. The system at present is very much not there to support home owners or home buyers. But those of you about to build or already building a house can almost immediately help make building your house a lot fairer than it otherwise would have been... and by doing this, you could kick-start the industry into stopping most defects from ever occurring. So get a capable building consultant before you sign, fight for your rights, and keep your resolve. If defects are 'nipped in the bud' during house construction, it should actually be cheaper for your builder than having a dispute at the end of the job.

With a proper definition of defect in place, perhaps one day a sufficient number of home owners of 'lemons' will fight all the way to a hearing, and be awarded sufficient amounts to rectify their homes properly... thus making negating and bluffing a thing of the past.

Chapter 12

When in Dispute, Ask the Wrong People and You Will Get the Wrong Answer

When your builder is very unreasonable or is refusing to rectify items you claim are defects, the time has possibly come to give up or to take an action against your builder. If you give up, I hope it's just over a few minor items; otherwise you have let the system bully you into submission.

If you give up for long enough, (say until the warranty ends), you may well find out that there were 10 times the value of defects that you thought existed in your home... and then you may well wish you had not taken the easier path of backing off.

We Australians got where we are today by fighting for our rights. Eventually it pays each and every one of us to stand up for ours. And we will need to continue to stand up for our rights if we want Australia to remain a fair place to live in. But nobody except a bully wants to fight if there is a reasonable alternative.

Well there seems to be no reasonable alternative as our system stands. Adjudicators and mediators and arbitrators must have a fair and adequate definition of defect or they will be on the side of the builder by default.

Many builders bluff, delay, refuse to answer your letter, band-aid most of what they agree to fix, and generally refuse to fix 90% of the defects including all of the expensive ones in your home. Many choose to look as if they care, but merely band-aid those items that will keep you happy so long as they can do them cheaply. Some of these 'repairs' will actually fail again even before the warranty period runs out.

So if the advice you have received seems unreasonable to you now that you know a little more about our system by reading to this point, seek other advice from capable people who actually know... and who actually care about your house.

If you are inquisitive enough to at least think about taking an action, then the first step is nearly always to contact a capable building consultant to find out all of the defects in your house including those items you asked the builder to fix. Of course your consultant would need to know what a defect is... and must be prepared to get in and on the roof and under the floor if your house has a timber floor frame.

Some of the defect items that you demanded be fixed by the builder may be ruled out as not claimable, because of what is contained in the former Building Commission document called "Guide to Standards and Tolerances. But there are many more that you will not know about. And these are usually more costly to repair... they are the 'hidden' defects. Some of these are difficult to reach and some are concealed below ground or beneath linings or in the (now) inaccessible framing; but there are sometimes symptoms that reveal these hidden defects and you may have a photograph that reveals some framing defects. What is needed is considerable thought and a fair bit of experience never hurts.

If you are thinking about taking an action to VCAT (or similar forum in states other than Victoria), you will need a capable team in order to get sufficient knowledge as to when and how to proceed. Maybe you will be lucky and the builder will be quite reasonable. But this is not likely at all, based on my experiences to date. In more than 90% of disputes I've come across there was no indication of reasonableness.

You may realise by now that a lack of a definition for the word defect has been responsible for a great number of problems, including a lot of home owners giving up in a dispute.

There are many people who when asking about their potential disputes, are persuaded to give up after they contacted the BCAV (Building Conciliatory Advice Victoria)… to the tune of well over 80% just on the telephone I believe). I heard this in a Forum on Building Construction during the days of the HGFL (Housing Guarantee Fund Limited), the only insurer at the time, and now disbanded.

If you used a building consultant who had no definition of defect, then I say that it is likely that the advice you obtained from BCAV may have saved you quite a bit of money.

Building Defect

But if you get a capable building consultant, you will be advised to ignore a great deal of advice received from the successors (VBA and repressive DBDRV in Victoria) to the ex-Building Commission or the BCAV or their building consultant service unless your questions are spot-on. The building consultants recommended by the authorities will probably look only at your list of defects too (as I believe they are obliged to do) AND they will inspect without any definition of defect. They will therefore not look at what you claim in the same light as a capable building consultant with a definition, because they are looking (I say) through special lenses where there has to be actual failure or imminent obvious failure or grossly unworkmanlike work with deformities exceeding those stated in the Guide to Standards and Tolerances. They will not look for any of the other defects in your house (defects that can be worth as much and even more than 10 times as much as those you discovered) and they are unlikely to consider the various reasonable life expectancies of the various components and systems that were used in your house.

If every home owner reads his/her contract prior to signing, and insists on a definition of defect and an upgraded definition of specification, houses might be well supervised once again, and eventually many building consultants and building case solicitors and barristers may need to look elsewhere for a living.

If the builder refuses to be fair, at least, BEFORE YOU SIGN, read (and understand) all the notes on your drawings, your specification, and your soil report and your contract so there are no unknowns.

Chapter 13

Building Contracts Need Updating Urgently

Definitions are a vital part of any standard. Without them a standard is little better than gobbledygook.

When there is a requirement such as at clauses 37 and 39 (regarding listing defects and defects that must be fixed by the builder) as in the 1997 privatized Domestic Building Industry HIA Contracts, then surely it is imperative that defect is actually defined therein.

Well it's not there... and it needs to be. Clause 11 of the same document goes part-way by describing what the builder shall warrant... but this does not go far enough as I have already pointed out. Clause 11.6 at least states something about being fit for purpose, but only regarding specific additional requirements of the owners... but the Building Act mentions compliance with the Building Code of Australia, which in turn places a mantle over the entire house that materials, components and systems will be installed so as to be fit for purpose. It would be nice if that implied warranty was actually mentioned specifically.

One saving grace however is the warranty that work shall be carried out in a proper and *'workmanlike'* manner. Now there's something to work with... except that workmanlike is not defined either... and *'specification'* IS defined, but so badly that each job has only half of what a specification actually is. It's pathetic! It leaves itself wide open to distortion... which builders have been only too willing to do.

So there is no definition of defect in the standard contracts for houses. How poor is that?

Even when something is obviously badly built, there is not much value in talking about that work (without a proper definition that each side understands), when the builder (temporarily at least) will be able to bluff the owner, merely by stating that the alleged items are commonplace and therefore not defects, thus (temporarily) getting out of the expense of rectifying the badly built item(s), and not delaying other jobs already underway.

This is the way many disputes begin. It's very difficult when the only expert (the builder you employed to build your dream home) tells you outright untruths.

And even when the builder states that he will attend to certain alleged defects, there is still a need to write a scope of works for every item to be rectified; or cheating via band-aids will undoubtedly occur, as it has to date since privatizing building surveyors began.

Rarely does the builder write a scope of works for each item to be rectified. And in my experience, this is because the builder is usually determined to do the absolute minimum of work possible to get the owner off his back. And it is all done under the name maintenance.

MAINTENANCE IN RELATION TO DISPUTES IS HARDLY EVER RECTIFICATION.

Specifications generally fail to inform home owners about workmanship... and the resulting watered-down documents are a major cause of the rapid decline in quality in my opinion. People are forgetting what good workmanship actually entails, if they ever knew.

If you buy a block of land, then get architectural plans drawn up, structural computations done, specifications written, a soil report carried out and then get a builder, sign a contract & obtain insurance and a building permit, (and obtain the finance of course); then you're ready to rock and roll.

Your house project can get underway.

These are your basic building blocks, your foundation stones, so to speak.

But the contract you signed contains the wrong definition of specification and no definition of defect at all? And it also states that the specification takes precedence over the plans.

So can you see why you need to read everything, before signing your contract?

Because I've seen some very interesting clauses slipped into specifications that limit the rights of the owners considerably.

One such clause permitted the builder to alter even the looks of the house without any right of challenging those changes. The Victorian look intended and drawn became the bell-cast look.

Often I've seen clauses slipped in that can permit the builder to alter stumps from concrete to cypress pine. Cypress pine is lighter but eventually rots on average in under 50 years.

I'm certain that many more similar clauses have been included in a large number of contracts.

So read the contract AND specification very carefully before you sign... and try to get some standard workmanship clauses included in the specification.

Quite commonplace are works to be carried out by owners; notes that are often added to drawings but not necessarily discussed with the owners so that they understand all of the implications... notes such as 'Agricultural drains and silt pits by owners'. If the silt pit is connected to the stormwater system as it normally is, how can the owner set the level of this drain without coordinating with the builder's plumber? But it clearly states in the contract that the owners or their agent must not give directions to the builder's sub-contractors, which makes it very difficult to lay the silt pits for the agricultural drains before the stormwater drains are installed.

So read the drawings very carefully before you sign.

The soil report often contains recommendations (that I maintain should OFTEN be listed as requirements)... and these recommendations are often ignored and I think not even read by some builders who think they know more than a soil engineer about the soil and its condition deep down.

So read the soil report carefully before you sign.

Or obtain the services of a capable building consultant before you sign, to point out the pitfalls in your documentation. Sometimes clauses can also be added to give you more rights, but these clauses are often shunned by the builders, even though they are very fair. They may also put the price up a little; but in my opinion they are probably worth it, unless the price rise is substantial. But it should not be substantial and that may give you a hint.

******

NOW, the word specification means 'specific description of' your building works and is defined in the dictionary as *'a detailed description of the dimensions, construction, workmanship, materials, etc., of work... to be done, prepared by an architect, engineer, etc.'*

Note the word workmanship!

I was taught that building specifications consist of 5 basic elements:

1. General conditions - including preliminaries, etc.
2. Codes to be complied with and pertinent to the works
3. Materials to be used in the construction
4. Workmanship requirements
5. Special conditions – for specialist works (out of the ordinary works)

Building Defect

Of late only an abbreviated Section1 and Section 3 appear in the majority of so-called specifications. It made me wonder. Why was this happening?

Part of the reason was that the building contract definition does not specifically mention workmanship (the result of labour) at all. So why should the builders go to the trouble of including these sections if nobody insists on them? So they don't... and I say it dumbs down the industry. What's more it keeps home owners from being 'in the know', so that they can't be nuisances by reading the labour component of the work and noticing that this and that were never carried out by the sub-contractors at all.

If the builder or building supervisor in charge of the work does not take the time to adequately supervise the project (with experienced eyes), then it permits short-cutting to take place, particularly if the payment for that work has been bargained down.

It is my firm opinion that 5 jobs not close to each other or perhaps as many as 8 jobs in say a unit development or in very close proximity to one another is more than enough for even an experienced builder to supervise well; let alone a supervisor experienced in just one trade.

And by the way, adequate supervision is the duty of every builder.

You've been charged for it...so make certain it happens. (Any slip-up is basically negligence on the builder's part).

Your (say) HIA contract defines specifications as *'the contract document that shows the full details of the*

Building Defect

building works and includes the details of the materials to be used.'

It is obvious that the word *'details'* has been interpreted to mean a general description of component parts of the project... such as "the works comprise the erection of a two storeyed rendered brick veneered cement tiled residence on a concrete slab with verandahs all round, a front fence and a cabana".

But there is rarely mention of labour at all; and labour amounts to about 45% of the work.

Full details indeed!

THERE IS NOTHING ABOUT WORKMANSHIP AT ALL.

And yet...

THE BUILDER WARRANTS THAT THE WORK SHALL BE CARRIED OUT IN A WORKMANLIKE MANNER

Talk about a farcical situation.

SO next is a short story on the history of specifications for just one trade (Painting) that shows why workmanship needs to be fully described... and why getting rid of all the rules and regulations leads to very poor quality work and general blight in the building industry.

Chapter 14

Paint Tin. Why Specifications are So Inadequate Today. Does Anyone Care?

Here's a little story on how to pass the buck – and a very appropriate story, because it was just about the first example in the building industry where all of the manufacturers (in the paint industry), together with many of the authorities and organizations, decided to take very little responsibility whatsoever for rotting timber. And from then on, it's been virtual anarchy.

Now it is fairly well known still, that red lead primer was probably the best means in the old days of preventing rot in timber. But lead paints were found to be quite dangerous to people when ingested. So they were eventually banned - I think in the early 50's... some considerable time ago anyway. (Take care all of you when renovating older homes).

So out came oil-based primer – usually pink... probably to differentiate between undercoat and primer.

This paint, (if you wish to use it on your house today and do a proper job of protecting your oregon beams and posts, hardwood framing or trims and white pine weatherboards, or any other non-durable exposed timberwork), has the special property of holding as a complete weatherproof film over splits in the timber. (A chemist expert in one leading paint company told me of this years ago when I started researching paints for the township of Rawson).

Timber can swell and shrink with variations in water content and in the heat can start to split. And when it splits, tiny rain-borne mould spores can (and do) lodge in these thin cracks and start to grow... and multiply... and grow... and multiply... and so on, until the piece of timber around the split is rotted. All mould needs to begin is water and a few nutrients, the smallest amount will often do and timber plus dust are quite adequate thanks very much.

So, the basic general specifications (the workmanship clauses) at the back of all of the...

- Commonwealth Savings Bank Specifications
- Housing Industry Association Standard Specifications
- Master Builders Association of Victoria Standard Specifications

as adopted by the Royal Australian Institute of Architects (Vic. Chapter) Housing Service... became the norm in the industry. Right through to today, the builder and architect associations still lend their names to the several times revised HIA & MBA specifications.

And almost at the very back of (say) the HIA booklets until at least 1985, was the painting and finishing clause that stated that for exterior finishes...

PRIMING – Door and window frames and sashes including glazing rebates in sashes and glazed doors, fascias, mouldings and other exposed timber shall be primed all round. Particular attention shall be paid to priming of end grain before fixing and whilst being fixed.

Note the repeated use of the word "shall".

Then in about the mid to late 80's Project Specifications came into being – very similar in appearance and scope at first glance.

However, upon closer scrutiny of these specifications, there at the end of the workmanship clauses was a revised painting section (as in the 1997 edition for instance), which read as follows…

11. PAINTING

11.1 Generally All painting is to be carried out … in accordance with recognised good trade practice and paint manufacturers recommendations.

1.4 Exterior Finishes
Painted Timber Surfaces Shall be given one coat of primer, one coat undercoat and one coat of exterior gloss enamel or alternatively, two coats of exterior acrylic paint.

AND THAT WAS ALL IT TOOK.

All of a sudden there was no mention of particular care to prime end grain whilst priming; just the general requirement to carry out work in accordance with good trade practice and to paint manufacturers recommendations.

(By the way it slows down carpenters considerably to paint end grain whilst framing, I can assure you).

So from about 1986 onwards, perhaps a little earlier, non-durable timbers such as oregon and dressed kiln-dried hardwood and white pine (sometimes referred to as Baltic pine), were installed with no end-grain oil-based primer protection. Most jobs I inspect reveal the total lack of use of oil-based primer, to end grain at least, from this date onwards. Sometimes windows and external doors are installed with inadequate protection as well. Sometimes the temporary protection on many timber windows and door frames and fascias and weatherboards requires an additional primer, but sadly this was and still is seldom carried out. Osmose could tell you all about it. I believe they make the temporary protective coating Protim used on windows and doors, mentioned earlier in the book.

So pergola beams, posts, fascias, gable end scotias, weatherboard corner stops, weatherboards (although often primed to the exposed face), cover straps, rails and quite a few windows were all of a sudden un-protected where it mattered most – at the end grain and at junctions.

So the rot began almost immediately, and from about 1990 the incidence of noticeable rot rose rapidly and continues through to today.

IT WAS BECOMING CLEAR THAT NOBODY REALLY CARED ANY MORE.

Once the specific clause requiring special care was removed, workmanship deteriorated rapidly. The authorities should have woken up to what happens when the rules are relaxed from just this one example. But instead, the freeing up of rules and regulations really got going big time, particularly after 1994. From 1997 onwards, it was at full throttle. In the mid-to-late *teens,* many of the codes and CSIRO (research authority) standards were to be made optional or superseded as well... sometimes with dire consequences (I say) as in the case of the CSIRO pamphlet 10-91 once referred to in most soil reports. There was very little if any outcry at the time from memory... to the detriment of many a house footing system.

I certainly don't remember any of the earlier mentioned organizations kicking up a storm of protest about the lack of oil-based priming either.

In fact the ex-Building Commission (and its successor the Victorian Building Authority) regarded paint defects as normal wear and tear instead of applying a reasonable life expectancy to this element... possibly because painting is the very first thing that owners notice and that would be a drag for builders and VCAT, (with so many poor paint jobs being done particularly to the outsides of houses where paint often powders off in under 2 years; often because no primer was used... or perhaps because the acrylic paint was applied to surfaces well below the required minimum temperature stated on the can (used to be a minimum of 10 degrees Celsius I believe). I've proven this to be the case on a few roof painting jobs and saw a painter painting weatherboards when the temperature was still 7 degrees C on a final inspection for one of my clients... and the overnight low temperature was just 4 degrees too.

And I don't seem to remember any paint manufacturers kicking up a storm even though the ex-Building Commission virtually downgraded the life expectancy of paint in one fell swoop. Perhaps they reasoned that everyone still wants their homes painted so it wouldn't make much difference.

So you may be getting the picture, and now understand why I cannot trust the authorities to really care about houses any more.

They have not truly cared for such a long time now, that not a lot of tradespeople seem to know what workmanlike means any more. The residential building industry as a whole needs people who are trained to think outside the square to be placed in charge of our standards, so that the homes constructed actually have a chance of being well built. People who knew what a good painting job entailed back in the early 80's are probably retired or close to retired by now: and with them goes all their workmanship expertise and all that care. I wonder how the industry will cope from now on... without proper specifications.

A whole generation has been hoodwinked into thinking that painting and rotting timber are purely maintenance items, and many do not seem to know that preventative measures can be taken to preserve even non-durable timbers against rot. The current generation of owners just fork out the money to handymen & painters to repeat the dose of non-durable timber / no oil-based primer / just use acrylic paint;

in just the way builders band-aid maintain their client's houses to get them over the (end of warranty) line.

BUT THIS IS NOT BUILDING *FIT FOR PURPOSE* AT ALL... IT'S THE EXACT OPPOSITE!!

It is stated in the industry by the recognized authority - the National Australian Forest Industry (NAFI) – that structures and claddings should be able to last at least 25 years, and structural elements 50 years. Now I don't know about you, but recently, I have rarely seen a set of 15 year-old oregon posts with no rot at the bottom. But I have seen a few older ones that were not rotted at 40 years of age... and these older ones were admittedly made of very old denser timber but also had end-grain oil-based primer applied including in the bolt-holes... and in 1962 oregon was classed by the CSIRO as being poor durability timber just as were most types of hardwood.

Surely this is all the more reason to require that the old rules stay in place for the lesser quality (faster growing) more open grained timber available today.

NOW that covers 'good trade practice'... so...

What about 'paint manufacturer recommendations?'

Where do I start?

Paint companies used to have and probably still have specification data sheets on all of their products (they're more likely to be on a CD or something nowadays). Well, I have looked up two leading brand data sheets on their oil-

Building Defect

based primers and the data sheets on their acrylic exterior paints.

And there is no mention whatsoever of durability required for timbers used outside; and no mention whatsoever on the unsuitability of acrylic paints used to protect some timbers outside, where those timbers are not first oil-base prime protected.

So that just leaves the paint tins to discuss... hence the chapter heading:

- The Oil-Based Primer tins (in effect) say that this paint is suitable for use on exterior raw timber - new, bare and burnt-off surfaces.
- The Exterior Acrylic Primer tins (in effect) say that this paint is suitable for use on exterior or interior bare timber surfaces. Sounds similar do you think?
- The Acrylic Top Coat tins (in effect) say that this paint is suitable for use on previously painted sanded acrylic paint surfaces and atop Exterior Acrylic Primer

As you can see there is not one mention of durability of the timber to be painted or suitability of the completed acrylic systems when painted on non-durable timber such as those mentioned above or worse still radiata pine being used outside more and more since about 1997. (Even LOSP treated pine is not totally durable... but that is not advertised).

THIS HAS GOT TO BE TOTAL DISREGARD OF RESPONSIBILITY REGARDING ROT BY ALL OF

THE AUTHORITIES. And it shows that less and less people care… and why so few know any more.

And it is totally and utterly unforgivable. All of the people involved should hang their collective heads in shame (and that would be a lot of hanging of heads).

******

Once exterior exposed raw non-durable timber components are fixed in place there is not much the painter can do in the way of priming end grain. So it's mostly not the painter's fault at all, although many people blame just the painters for rotting timbers.

And if you have primed and painted the surfaces but not the end-grain, the water can be held in for longer periods of time, sometimes causing rot faster than not painting at all. The painter can only paint what is accessible. (I have seen a case where the unpainted west facing oregon fascias on the garage of one house lasted longer than the painted west facing oregon fascias of the identical house next door and built at the same time by the same builder).

The fault rests partly with the manufacturer, partly with the specification writer, partly with the builder & his carpenter, and partly with the building inspector who should not have passed it in my opinion, particularly where involving structural components. Because many exposed structural beams today are simply not durable… and quite a few are rotting.

Often contracts state that *'painting is to be by owners'* to lower the contract price. But in my opinion, this exclusion from the contract does not exempt the builder from rot-related problems during the 6.5 years warranty period, because the end grain was not made accessible for the owner to prime prior to erection of the exposed beams or weatherboards.

By the way, most painters still somehow know what a good paint job entails. It's just that so few people (and hardly any builders) ask for a proper job when told the price... but if you want a long-lasting job and no costly rot repairs after just 4 years or so...

DON'T USE NON-DURABLE TIMBER OUTSIDE IN EXPOSED SITUATIONS.

That's what I say. But the government will not regulate this problem. Perhaps this is because some industries (enterprises) will be affected (just as for red gum stumps and cypress fence posts with no tanking), particularly if oil-based priming rules are not reinstated. So on it goes unchecked... and nobody has to own up (or has owned up) to having made these mistakes...

And the cost of new houses is cheaper... in the short-term.

BUT IT COSTS FAR MORE IN THE LONG TERM IF NON-DURABLE TIMBER IS NOT PRIMED.

Eventually lack of priming costs the industry considerably more because of the replacement of rotted timbers. And governments still think that because the cost of housing will increase, fussy painting rules and restrictive codes will be bad for the economy. I hope you can now see how poor

Building Defect

that thinking is. This typical example is just 3 years old and just about to rot.

And this blinkered approach is being repeated right across all areas of the residential building industry as governments see the price of housing increasing faster than inflation.

Building Defect

This was a job where painting was to be by the owner. How is the owner supposed to protect this hardwood lintel at its ends and at the post joints? And look at the pathetic attempt to match the height of the other beams with little hardwood in-fill pieces under the lintel all without oil-based primer and with a multitude of joints to allow water entry... and what's the bet that the yellow primed timber members were also not end-grain treated?

But higher costs for more durable (and virtually defect-free) houses just means that houses need to be a little bit smaller. I think we can all afford that small decrease in the size of our houses, which have nearly doubled in size in the last 30 years, with little if any increase in the size of families.

******

The answers I received from the painters at two different properties adjacent those I was inspecting were very illuminating. I was able to ask the identical question – "Do you know what timber that is that you are painting?" (Whilst pointing at the timber slats on the gable ends they were about to paint /actually painting). The replies were almost identical – hardwood I think. And I found out that each had been told to apply just one coat of acrylic paint or whatever it takes to make it look good... and to hurry up so that the builder could sell the units. (One is photographed at just 9 years old below)

And I thought it didn't really matter much if they painted them or not... one coat or two coats of acrylic... either way, they were likely to rot in just a few years... because...

An all-acrylic paint system alone is not suitable for exposed non-durable timber and doesn't adequately protect against rot. Basically it is not fit for purpose.

******

And so there are about enormous numbers of houses, units and additions in Victoria built since 1985, with non-durable timber components in exposed situations, not protected in accordance with that poor durability... and there are possibly even more repairs done without the oil based primer too. So we need to fight to improve our specifications.

I believe that the Victorian government either has, or is currently in the middle of, grading Australian Codes according to their relevance. Seldom-used codes are virtually going to become extinct I believe. And it may also eventuate that if such codes are not specified then they cannot later be used to prove that works are defective.

Building Defect

To me all of this appears to be governments continuing to blunder on, removing as many restrictions as they can in order to keep the price of housing down... without considering how a definition of defect will achieve a far better result overall.

With maintenance costs substantially reduced by the small increase in the cost of acquiring experienced supervision and the payment to tradesmen t avoid short-cuts, the overall price would still be reduced... **BUT the economy would ALSO be a whole lot better off, by owners not having to continue the maintenance of their short-cut riddled houses.**

Otherwise the eventual outcome will be a further lowering of the quality of houses being built.

Chapter 15

What You Can Do to Give Your New Home a Fighting Chance of Being Built Properly

Everyone can do quite a few things to ensure that their houses are built well.

Firstly, get a capable builder who fully supervises his own work.

OR

Get a capable building consultant to peruse the drawings, specification and contract before you sign the contract, or failing that, at least before construction begins.

AND

Get that capable building consultant who defines defect to inspect the works at each stage of construction. But do not employ a building consultant with lots of disclaimers.

******

One (or preferably more) of the following avenues may help considerably in acquiring a competent builder for your new home or new additions:

- Talk to home owners who built with your builder

Building Defect

- Ask to see a house your proposed builder recently built.
- Ensure the builder is a registered (domestic builder - unlimited) building practitioner.
- Ensure that the builder is currently insured.
- Try to ensure that the person signing the contract is a registered (DBU) building practitioner and try to insist that this person actually supervises your job - I suggest at each stage OR... find out if the company is going to use a supervisor who is a registered building practitioner and if not, just how the company is going to ensure that the job is carried out in a workmanlike manner. This should be in writing and in the contract. See if you can contract that adequate supervision shall be the responsibility of the registered DBU unlimited builder in the company... good luck with this very important clause!
- Ensure that the builder takes out building warranty insurance in his or his company's name (but exactly the same name as in the contract).
- If the builder prints the contract and it is not a coloured booklet printed by the builder's own housing authority, but merely photocopies, ensure that the builder has written permission to print such contracts. Then ensure that the contract is not a superseded model.
- Ensure that you read and understand the contract before you sign. (A discussion with your capable building consultant or solicitor regarding the schedules in your contracts can also prove very beneficial in some instances. Matters such as fair penalties for late completion caused by the builder, prime cost items and similar matters).
- Ensure that your contract spells out what the various stages entail in detail, and try to ensure that all sections of the building are required to have progressed to the same extent at each stage, before the builder claims payment.

- Ensure that you and your building consultant, (if you choose to employ one), is permitted to visit the site and inspect at any / each stage (by appointment naturally)
- Ensure that you are to be given a FULL set of stamped building permit documents including those stamped after construction begins (such as truss layout plan and computations)
- Ensure that specification enhancement clauses are included in the contract (or send these to the builder after signing and state that the contents of those clauses are going to form the basis of the inspections carried out for you). These clauses should address the issue of alternative solutions being pre-approved by you the home owners, or that they are not permitted at all. Good luck with this one too.

Finally, (and this may also be difficult), I suggest that you…

- Try to insist that you choose the building surveyor (it can be the one the builder normally uses), but then let the building surveyor and the building inspector know that they are carrying out the work for you and not the builder. In particular I suggest, (after you have paid), that you write to the building surveyor that you expect the inspections at each stage to reveal all of the breaches of the building code deemed-to-comply provisions in writing to you (best if a copy is also sent to the builder so as not to delay the builder).

There are about 30 or so general workmanship clauses (different for every job unfortunately), that would also substantially help owners who are having their house built for them… and they're not unreasonable in any way whatsoever. So the price should not alter, and if it does then be very wary of that builder I suggest.

There are also special clauses that can relate to matters such as... fall in land allowance (make certain that the builder did the levels or checked the levels drawn before signing), soil classification, substitutions, areas of work by the owner (quite often not realized by the owners)... and even the type of stump to be used. "Painting by owners" and "agricultural drainage by owners" for instance need qualification or you are letting yourselves in for built-in defects before you even start. 'Driveway by owners' is another to be very wary of particularly if the levels are tight. Many owners' cars scrape on defective driveways.

Try to have the drawings take precedence over the specification (but only if you have the drawings checked by a capable building consultant (architects and building surveyors are probably best at this). Or at least have the specification perused by your capable building consultant *with a fine toothcomb*. Because in the contract at present, the specification takes precedence over the drawings... and if the builder wishes, he could state in the specification anything he wants to... items such as 'the house is to be reduced by 45 square metres and to the builder's design as necessary'; or 'the spa bath, front porch, drainage and retaining walls are to be done by the owner or at the owner's extra expense'... whatever he wants to incorporate basically... SO READ IT. Fortunately he has to give you a copy. Be certain to read it very carefully indeed... **Then sign.**

'The builder shall, prior to signing, verify all levels and dimensions on site: and the agreed price includes the construction of the house as drawn and in accordance with those levels and the soil report'... is the sort of vital clause that should appear in every contract in my opinion. There are several more that would also help.

Building Inspectors seem to exercise far less control over builders than they did in the 80's.

Each builder has usually chosen his own Building Surveyor since Building Surveyors and Building Inspectors were privatised, and a dependency of each on the other may have led to considerable bending of the rules... past what is reasonable. It would explain a lot.

I can think of no other reason that explains why there are so many defects in the building work at completion time except a general lack of caring. It has to be either a type of cronyism or a misunderstanding of basic terms such as workmanlike, alternative solution and performance by building inspectors.

Be pro-active while you still have some control. Later on it may simply be too late.

Do not sign until you have an excellent chance of a good finished product. You may need a bit of luck in your choice of builder and/or building consultant... but this book should help considerably.

To all of you home owners, particularly if you are likely to go with the builder with the lowest quote, I strongly suggest that you do at least the following prior to signing: -

Upgrade your documents to:

- Redefine specification,
- Remove unfair builder clauses,
- Up-grade the builder's specification by adding workmanship clauses
- Ensure that good quality standards are clearly defined as not permitting alternative solutions based on expert judgement without your agreement as to the short-changing, and that The Building Code of Australia (2000 edition rather than the latest edition) deemed-to-comply provisions apply unless otherwise agreed to by you..
- Insist that manufacturer, structural and soil report recommendations become requirements,

And best of luck if there is no agreed public definition of defect for the whole industry.

And if there is none when you are about to build, then I suggest you use the definition in this book and spell out fully what specification means.

Chapter 16

What Needs to be Done RIGHT NOW to Stop Blight in the Domestic Building Industry

It would be very beneficial to the building industry in the longer-term, if building consultants (present and future) were made to pass a series of stringent tests in order to become registered 'opinion writers' as a new class of building practitioner over, say, a 3- year period, having to prove that they have sufficient knowledge and/or experience to hold an opinion.

These people I believe would then be more or less equivalent to what are called Surveyors in England. In England, as I believe may have been the case in Canberra, pre-purchase inspections are (were) mandatory in the sale of a house. So why not make these compulsory at the completion of a new house or additions, once we have a definition of defect?

These tests shouldn't preclude new-comers from becoming registered or those already in the system from continuing, as long as they prove they have what it takes to inspect houses, either from knowledge or experience or both: plus for new houses they should know what a defect is and what adequate performance entails.

Building Defect

I've given a fair bit of thought to this topic, in order to attempt to stop the blight so noticeable in the houses churned out since 1996. But certain government decisions mentioned throughout the book could put a huge obstacle in the way of this advice.

To achieve a return to well-built houses, several basic building blocks need to be up-graded significantly as follows:

- We need to define adequately the word DEFECT
- We need to expand the definition of the word SPECIFICATION to include labour
- We need to stop warping the words PERFORMANCE and PERFORM
- Experts such as soil reporters and manufacturers should be made to STOP RECOMMENDING and START REQUIRING that certain precautionary measures be undertaken

And of course:

- Building Contracts will need to be upgraded
- The Code for Inspection of Buildings needs to be re-written to include all houses including new houses and the disclaimers to do with reasonable access re-examined PLUS include a full definition of defect
- The Building Code of Australia needs to be upgraded with the reinstatement of several omitted deemed-to-comply provisions to avid short-cutting
- New authorities should be formed to oversee most trades to cut down short-cutting

Some other measures that would improve the situation are:

- Replacement of the unfair warranty insurance in place at the moment, even if it has to cost more. (The insuring of all defects for 6 years is the main suggested upgrade but not the only one).
- Building consultants should have to be insured for their inspection (opinion) reports
- Building inspectors should be paid more for more thorough inspections and held responsible by an authority for their lack of discovering short-cuts and defects
- Building supervisors should be required to be the equivalent of unlimited responsibility builders... with I suggest a three year minimum apprenticeship
- Sub-contractors could be asked to sign contracts with builders and required to give certificates of compliance for the work they have carried out
- The Building Practitioners Board, at least in the short term, should carry out spot inspections of a large sample of houses to ensure more quality control; with repeat offenders at least temporarily stopped from trading or placed on probation.
- A register of commonplace failed systems needs to be built up, at least in the short-term with a view to setting up a regulatory framework to safeguard the longevity of buildings and their component parts.
- Legal opinion in building consultant reports should be stopped.
- Compulsory education of building consultants and mediators, particularly on basic terminology should be instigated using an official definition of defect.

- Alterations to the mediation process just for the short-term, to sort out the differences of expert opinion by a board of capable building consultants, prior to the first mediation, using a proper definition of defect.
- Remove the current indemnification of building authority-appointed building consultants so that they must use the proper definition of defect also.
- Prohibit the recommending of experts by building authorities.

We still depend on a ponderous legal system that is heavily dependent on precedent, and yet the combined expert knowledge in codes is challenged and pushed aside without sufficient countering measures to limit the so-called *expert judgement* used in commonplace short-cutting and the vast majority of work that masquerades as acceptable *alternative solutions*.

The courts need to be challenged as to onus of proof and just what is expert judgement; to sort out thoroughly issues such as: **fit for purpose / likely / reasonable life expectancy of proposed rectifications** and the allowing of words such as **defect / maintenance / rectify** to go unchallenged.

The exclusion from hearings of some building consultants who may well be more knowledgeable about ground movements than some structural engineers and/or building surveyors (for instance) should also cease.

I think VCAT should be there to decide what is reasonable, because people on both sides can be unreasonable, but courts need to base their decisions on a solid foundation of well-defined building block terms and cogent standards for the sake of the houses.

If that means that the Building Act and the Contracts Guarantee Act need to be upgraded, then let's do it... as soon as possible... before the economy suffers from the blight (obvious to so few) setting into the residential building industry.

So by simply insisting on:

- A correct and full definition of defect.
- Adequate supervision of all new work by experienced *unlimited* builders or experienced building consultants or qualified supervisors.
- Enforced performance checks by building surveyors on the building inspectors they use and the acquiring of knowledge as to what defects and alternative solutions the building inspectors have been permitting (perhaps to start a register of what is not permitted)...

We will go a long way to reducing the likelihood of built-in defects and save ourselves a costly blight on the economy of the country... and actually reduce the need for disputes.

******

Then I think builders and tradespeople will realise that they have to improve their thoroughness and take total care once again, so that good quality houses will again be the order of the day. Then builders and tradespeople can all truly be proud of what they have achieved... and building consultants, building case solicitors & barristers / tribunals

Building Defect

may hardly be necessary any more. Wouldn't that be fantastic!

For the benefit of at least the residential building industry (just as it is with disease as regards the human population as a whole), we need to approach everything according to the wise saying by Erasmus, that:

Prevention is Better than Cure.

And that is what this book hopes to achieve…

        The prevention of a workmanship blight…

        For the sake of the residential building industry…

        For the sake of the economy as a whole…

For the good of all Australians now and into the future.

Building Defect

CPSIA information can be obtained
at www.ICGtesting.com
Printed in the USA
BVHW092338140223
658492BV00022B/888